A Guide to FIELD GEOLOGY

Dr. N.W. Gokhale
Retired Professor of Geology,
Shreyas, 82, Hubli Road,
Dharwad-580007, Karnataka

CBSPD

CBS Publishers & Distributors Pvt Ltd

New Delhi • Bengaluru • Chennai • Kochi • Kolkata • Lucknow • Mumbai

Hyderabad • Jharkhand • Nagpur • Patna • Pune • Uttarakhand

Disclaimer

Science and technology are constantly changing fields. New research and experience broaden the scope of information and knowledge. The author has tried his best in giving information available to him while preparing the material for this book. Although, all efforts have been made to ensure optimum accuracy of the material, yet it is quite possible some errors might have been left uncorrected. The publisher, the printer and the author will not be held responsible for any inadvertent errors, or inaccuracies.

A Guide to Field Geology

ISBN: 978-81-239-0749-9

First Edition: 2001

Reprint: 2006, 2007, 2009, 2010, 2011, 2012, 2015, 2018, 2023

Published by **Satish Kumar Jain** and produced by **Varun Jain** for
CBS Publishers & Distributors Pvt Ltd
4819/XI Prahlad Street, 24 Ansari Road, Daryaganj, New Delhi 110 002, India
Ph: 011-23289259, 23266861 Website: www.cbspd.com
e-mail: delhi@cbspd.com

Corporate Office: 204 FIE, Industrial Area, Patparganj, Delhi 110 092, India
Ph: 011-4934 4934 Fax: 011-4934 4935 e-mail: publishing@cbspd.com; publicity@cbspd.com

Branches

- **Bengaluru:** Seema House 2975, 17th Cross, KR Road, Banasankari 2nd Stage, Bengaluru 560 070, Karnataka, India
  Ph: +91-80-26771678/79 Fax: +91-80-26771680 e-mail: bangalore@cbspd.com
- **Chennai:** 7, Subbaraya Street, Shenoy Nagar, Chennai 600 030, Tamil Nadu, India
  Ph: +91-44-26680620, 26681266 Fax: +91-44-42032115 e-mail: chennai@cbspd.com
- **Kochi:** 42/1325, 1326, Power House Road, Opp KSEB, Power House, Ernakulum Kochi 682 018, Kerala, India
  Ph: +91-484-4059061-65,67 Fax: +91-484-4059065 e-mail: kochi@cbspd.com
- **Kolkata:** 147, Hind Ceramics Compound, 1st Floor, Nilgunj Road, Belghoria, Kolkata-700056, West Bengal, India
  Ph: +033-25633055, 033-25633056 e-mail: kolkata@cbspd.com
- **Lucknow:** Basement, Khushnuma Complex, 7 Meerabai Marg (Behind Jawahar Bhawan),Lucknow-226001, UP, India
  Ph: +91-522-4000032 e-mail: tiwari.lucknow@cbspd.com
- **Mumbai:** PWD Shed, Gala no 25/26, Ramchandra Bhatt Marg, Next to JJ Hospital Gate no. 2, Opp. Union Bank of India Noorbaug, Mumbai-400009, Maharashtra, India
  Ph: 022-66661880/89 e-mail: mumbai@cbspd.com

Representatives

- Hyderabad 0-9885175004
- Jharkhand 0-9811541605
- Nagpur 0-9421945513
- Patna 0-9334159340
- Pune 0-9923910676
- Uttarakhand 0-9716462459

Printed at: SRK Graphics, Shahdara, Delhi, India

FOREWORD

The early man on mother earth with his shaky steps in stone age, slowly leaped into great heights and space-age marked altogether new era in human development. In this crusade the contribution and role played by Geologists need to be understood and complimented by one and all.

The field geologists work for years and decades in the remotest areas for proving mineral deposits. Based on their work, industries and ancillaries come up and developmental activities are geared up. Field geology is a very important branch of geology. For field geologists the only textbook available was "Field Geology" by Frederic H. Lahee which remained the only companion to them for decade after decade.

In my opinion there was a long-felt need to have one such book on field geology. It is very heartening to see that an eminent professor from Dharwar University Dr. N.W. Gokhale has now brought out an excellent book, "A Guide to Field Geology". Dr. Gokhale by his painstaking efforts, with decades of experience in teaching and sound experience in field work could present this book to all of us, in his very lucid language and with vivid commentary. The nine chapters covering all parameters of field geology, incorporating 33 figures and appended by 15 photographs fulfil the long-felt requirement of a book on field geology in Indian context by an Indian Professor.

No doubt there may be several books on structural geology, however, I am sure, this book will be of immense use to the students of geology in India. Dr. Gokhale has dealt field geology in this book in such a way that it is also a valuable reference book to field geologists all over India and to mining engineer I feel even layman and trekkers can find number of useful tips so that they can enjoy their outing and collect rock/mineral samples with more clarity and knowledge. This book can definitely adorn personal library of all geoscientists in India.

Major part of my career was spent in the field, working in various geological formations and the need to have such a guidebook was every time felt. I am very happy to note that this long-standing requirement of field geologists has been taken care of to a greater degree.

I am very much thankful to Dr. N.W. Gokhale for giving me an opportunity to express my views on this book and also what I feel about it.

Nagpur

D.A. Bhagwatwar
Joint Director (Retired)
Directorate of Geology and Mining,
Government of Maharashtra, Nagpur

PREFACE

Treasures and the mysteries of the earth are buried in the rocks. Recognising the varieties of rocks, and their distribution in the continental parts of the crust of the earth, constitute the prime objective assignments for the geologists. Depending upon the accurate feedback from the said two parameters, the treasures and the mysteries of the earth, get resolved. Success therefore depends upon the ability of the geologist to identify the rocks in their natural habitat, and to prepare a documentation of their dispositions in the crust of the earth. "What is observed, and whether observed or not" by the geologists, formulates the basis of building up the three-dimensional model of the earth. The data collected during the field studies, are later subjected to the laboratorial studies. Therefore the validity of the outcome of the laboratorial investigations, entirely depends upon the accuracy of the "Field Data". The field studies thus acquire a profound importance in geology. Field mapping is not receiving sufficient attention, because of overshadowing by photogeology and the satellite imageries. However there is no alternative for the field work. Considering this background, an attempt is made to enlighten the budding geoscientists about the essential aspects of field geology.

The various components of field geology have been described under nine chapters. The text has been illustrated with 33 figures and 15 field photographs. The petrographical and structural studies, commence with the field mapping exercise. How to identify the rocks and the structural elements possessed by them, in the field, has been stressed and elaborated. One of the highlights of this book is the compilation of a "Tabular schematic field classification of the igneous rocks". These rocks have been given priority because, these are extensively developed in the crust of the earth, and there are innumerable varieties within them. A separate chapter is devoted to the "Field Indicators" of rocks, minerals, and structures. This has the semblance of a "Ready Reckoner".

The author is extremely grateful to Shri D.A. Bhagwatwar, Joint Director (Retired), Directorate of Geology and Mining, Government of Maharashtra, Nagpur, for his appreciatively and critically worded "Foreword" to this book. This has decidedly increased the academic value of this book, and its suitability to the field geologists. I am thankful to Doctors D.I. Deendar, H.D. Desai, N.N. Gothe, V.N. Hegde, G.V. Hegde, V.B. Koppad, G.S. Pujar, S.C. Puranik, D. Muralidharan, and B.P. Waghamare, for providing the field photographs, and figures from their doctoral theses, to me, to be incorporated in the present book. Some photographs are indeed rare (Brickwork pattern, Pillow structures, Orbicular structures). The field photographs will give an idea about the appearance of the rocks and certain peculiar structures, in their natural habitat. I am equally thankful to the CBS Publishers & Distributors, New Delhi, for composing, printing and publishing this book in an excellent fashion.

Dharwad 580007 (Dr. N.W. Gokhale)

CONTENTS

CHAPTER 1

Role of Field Studies

INTRODUCTION

Geologists, also known as Geoscientists, are trying to understand the constitution of the earth. The surficial portion, which is directly visible to the geologists, is made up more of water bodies (the seas, the oceans, and the lakes, together constitute nearly 2/3rds of the entire surface of the earth) than the land masses (Africa, America, Asia etc.). Large area of land around the north and the southpoles, is covered with a thick blanket of ice sheet, throughout the year. The higher peaks of the mountain ranges, are likewise covered by ice. The radius of the earth is 6400 km, but the exposed portion of the earth above the sea level, is a mere 8 km, which is available for the direct observations of the geologists. The deepest mine in the Kolar Gold Fields, Karnataka, India, is just about 3.2 km in depth. Therefore, it becomes inevitable to collect numerous samples of rocks, not only from the different parts of the surface of the earth, but also from depth (mines) too, in order to obtain as true a composition of the crustal rocks, as is possible.

It is absolutely necessary to locate the sites of collection of the samples of rocks, both "areawise and depthwise", towards the objective of establishing the constitution of the crust of the earth. Depending upon the accuracy of collection of the samples of the rocks, and their spacewise distribution, the geological maps prepared therefrom, will be reliable and indisputable ones. Validity of the further studies to be carried out in the laboratory, therefore, depends entirely upon a reliable FIELD DATA. Hence in the geological pursuits, the "Field Studies" gain extraordinary importance. The role of these studies, is akin to that of locating a suitable site for the "Foundation of dam across a river, foundation of the sky scraper building, or any other similar structure". A weak or a faulty foundation, eventually leads to the collapse of the structure built over it. Likewise, a "faulty or an incorrect field data" will culminate into an "incorrect model of the constitution of the crust of the earth". It is therefore necessary to understand the various components of a systematic and an all-proof "Field Studies".

COMPONENTS OF GEOLOGICAL INVESTIGATIONS

The geological investigations may be split into i) FIELD, and ii) LABORATORY, studies. Unlike the other branches of science (Physics, Chemistry, Biology etc.), experiments in the strict sense of the word, are not at all possible in geology. For example, it is not possible to prepare rocks in the four walls of any laboratory, on the scale, as are found in the crust of the earth. Geoscientists have advocated mechanisms like the Plate Tectonics, to explain the generation of magma, causes of earthquakes, and so on. However, the geological investigations are mostly of the nature of "identification"; whether of a "rock, a mineral, a fault, or a fossil, and so on". Ultimately, documenting the "spacewise distribution of the rocks, the structural features like the faults, the folds" in the form of a map, is the main aim of the field investigations. The "Laboratory Investigations" like the megascopic and the microscopic examinations, the chemical analysis of the samples of rocks" could be carried out, only if the field studies are undertaken. *The laboratory studies are to strengthen the observations made during the field studies, and not vice versa.* Since this book has been designed to describe the various aspects of "field geology", the "laboratory studies", have not been dealt with. It will not be an "over statement', if it is said that the "field studies" claim a lion's share in the "Geological Investigations".

Having realised the role played by the "field studies", let us understand the manner in which these studies are to be conducted. The various rocks and the other features encountered in the crust of the earth, are required to be documented in the form of a map. The surficial appearance (hills, valleys, rivers, streams etc.) of the earth, is presented as a "Topographical Map". These maps are constructed on different scales, depending upon the details required. On such maps, the geological data collected during the field work, are to be recorded. The topographical maps form the FIRST COMPONENT of the field studies. The other aspects of the topographical maps will be described in Chapter 2. The Geological Map constitutes the SECOND COMPONENT of the field studies. These maps show the rocks and the structural features possessed by them. The other details of this component, are described in Chapter 3. The crustal rocks belong to three categories, namely, the igneous, the sedimentary, and the metamorphic rocks, which possess different field features. Therefore the methods to be adopted in mapping them, also are different. Chapter 4 deals with the procedures followed in preparing maps pertaining to the three categories of rocks. A geologist also should be familiar with the salient petrographical features characterising the igneous, the sedimentary, and the metamorphic rocks, which build up the crust of the earth. Otherwise he will not be in a position to distinguish one category of rock from the other, and in turn, he will not be collecting samples, believing that a different rock has not cropped up, and it is the same one as was collected from the previous outcrop. The details asto how the distinctions (petrographical) are to be made in the field, are described in Chapter 5.

As said earlier, the igneous, the sedimentary, and the metamorphic rocks, possess different structural elements. These are also required to be observed and shown on the map. The different structural elements have been described in Chapter 6. Ultimately, the field geologist desires to know the nature of the data to be collected during the field mapping exercise. In the case of the igneous rocks, the emphasis is required to be laid more on the petrographical features, besides those that are useful in proving the instusive or the extrusive nature of the rock body. With the sedimentary rocks, the primary structural features like the current bedding, the ripples marks etc., assume importance. In the case of the metamorphic rocks, the geologist is interested in ascertaining the grade of metamorphism attained by the rocks. Therefore the type of emphasis to be laid during the collection of the data, varies, according to the category of the rocks, and the "collection of data" assumes paramount importance. The details of the collection of data, are set out in Chapter 7.

The most important and the crucial aspect of the "field studies", is to identify the rocks and the structures possessed by them. In order to assist the geologist in this regard, some indicators (reckoners) of rocks, minerals, and structures, have been enlisted in Chapter 8. The fact remains, that these indications should be strengthened or rectified by the detailed laboratory investigations, especially, the minerals and the rocks.

The geological map is a document displaying the disposition of the different rocks, as are occurring in the crust of the earth. The exposures of the rocks at the surface of the earth, are described by the term “outcrops”. It is necessary to remember that the crustal rocks continue at depth to a varying extent. It is therefore necessary to ascertain the extent to which the rocks are likely to continue at depth, from their behaviour at the surface of the earth. Though an exact prediction is generally not possible, at least some estimation could be made. The depth is deduced from the shapes of the rock unit (in the case of the igneous rocks), while the amount of dip, the nature of the folds, the amount of throw of the fault, and so on, are useful in the sedimentary rocks. The geological map, therefore, also should accommodate the structural features of the rocks. Compared to the surficial extent of the crustal rocks (i.e., the size of the outcrops in respect of its length and breadth), the size of the maps prepared, is considerably small. Various symbols are therefore made use of, in marking the important structural features on the map. Chapter 9 has enlisted symbols that are frequently used during the field mapping exercise. It has also enlisted the symbols used in the presentation of features like a fort, a temple, on the topographical maps.

SUMMARY

Field studies are extremely important. These form the basis for the later laboratory investigations. Incorrect and an incomplete field data, result in an erroneous assessment of the composition of the crustal rocks. The land surface is extremely small, as compared to that of the oceanic water. Large parts of the land surface, are covered by forests, soils. The outcrops of rocks are meagre. Therefore efforts should be made to study all the outcrops, during the field mapping exercise.

CHAPTER 2

Components of Field Studies

INTRODUCTION

The data collected during the field studies, are needed to be documented. The data are plotted on a topographical map, pertaining to the area undertaken for the geological mapping exercise. The topographical maps, therefore constitute the starting point of the field studies. These maps depict the surficial relief of the earth. It is important to remember, that, fundamentally, the hills, the valleys, the plane ground, are all made up of rocks. The geologist is inquisitive about them (rocks). It is therefore possible to get a broad and a general idea about the kind of the rocks, from the geomorphological features (like the hills, the valleys etc.). It is appropriate therefore, to devote a separate chapter to describe the correlation between the geomorphology and the crustal rocks. Besides the topographical maps, in the latter half of the 20th century, the aerial photographs and the satellite imageries, are being used extensively in the geological investigations. The information gathered from the aerial photographs, and the satellite imageries, also helps to identify the rocks exposed on the surface of the earth. The topographical maps, the aerial photographs, and the satellite imageries, constitute the FIRST COMPONENT of the field studies.

Besides the topographical maps, the aerial photographs, and the satellite imageries, the geologist needs a number instruments. No field work is possible without the proper accompaniments of hammers, clinometer compass, pocket lens, penknife, and so on. Out of these items, the compass, the pocket lens, and the hammers, are indispensable, because these constitute the most essential ones. All these have been therefore categorised as composing the SECOND COMPONENT of the field studies. In the following paragraphs, a detailed account of the two components, has been presented.

THE FIRST COMPONENT

The Topographical Maps

The Survey of India publishes these maps (these are also called as the Toposheets) on several scales. The earlier system of measurement was in inches, and miles, but currently it has been replaced by centimeters and kilometers. Several scales in use, are as under;

One inch = one mile, four miles, eight miles, 16 miles, 32 miles.
One inch = 330 feet, 660 feet, half a furlong, one furlong.
One centimetre = one km., 2.5 km., 10 km. 25 km.
One centimetre = 100 metres, 200 metres, 500 metres

Meaning of a map with a scale of "one inch equal to one mile" is that on such a map, a length of one inch on the map, corresponds to a distance of one mile on the ground. Depending upon the scale specified on the map, distances corresponding to a length of one inch could be 4, 8, 16, 32 miles on the ground. Likewise a length of one centimeter on the map (having a scale of one centimeter equal to one kilometer etc.), may correspond to a distance of 1,2.5, 5, 10 or 25 kilometers, and so on, on the ground.

The geological mapping is generally effected on bigger scales like 4 inches equal to one mile., or even on still bigger scales. A bigger scale is chosen in order to mark small sized items like the basic and the acidic dykes (these are hardly one to two meters in width), shear zones, trends of joints, coal seams, and so on, on the maps. Therefore depending upon the minute details that are required to be mapped, either the one inch toposheet, is enlarged to the required scale, and or separate "field sketches" are prepared on such scales as, one inch equal to 10 feet, one centimeter equal to 1 meter, or on still bigger scales. The point to be noted is that the size of the toposheet (map) is considerably small as compared to the area to be accommodated on it. Area of toposheet having a scale of one centimeter equal to 500 meters is only 0.003 km^2, whereas the area of land surface covered on it is 750 km^2.

The dimensions (length, breadth or width) apparently get shrunk, when plotted on the map. On such a map, a line of one millimeter length will correspond to 50 meters on the ground. Therefore anything less than 50 meters, cannot be shown at all, on such maps, because it will have an extent less than one millimeter. Such a marking could hardly be seen on the map. This aspect should not culminate into inaccuracy, while plotting the geological features on the topographical maps. It may therefore be said that "smaller the dimensions of the geological features, bigger should be the scale of the topographical map", and "bigger the dimensions of the geological features, smaller should be the scale of the topographical map".

THE AERIAL PHOTOGRAPHS AND THE SATELLITE IMAGERIES

The aerial photographs are similar to the black-white photographs, as far as the their appearance is concerned. The land surface is photographed from an aeroplane. Depending upon the intensity of the sunlight being reflected, and or refracted by the surface of the earth (rocks), the tone of the print (the aerial photograph) will vary. Analogy that can be given is of photographs taken of two persons- one having a dark complexion, and the other having a fair complexion. The black-white prints of these two persons are found to differ significantly, in the shades of gray. It is argued that the rocks differ in their intensity of reflection, and or refraction, when light rays impinge on them. Therefore from the tone of the aerial photographs, rocks could be distinguished from one another. Likewise, the geomorphic features (hills, valleys, gorges and so on) also could be read off, through a careful scrutiny of the aerial photographs. However these photographs have a major disadvantage. These do not have a scale, as well as a north line marked on them. The direction of north line is required to be derived from the direction of "flight line" of the aeroplane that had taken the photographs. The scale is to be established through a complicated procedure. Highly mountainous terraines, or thickly forested regions, become inaccessible for field work. Such areas could be conveniently aerially photographed, and the broad petrographical and the structural features could be read off. In addition, major structural features, like a graben a horst, qua-qua versal dip, major anticline

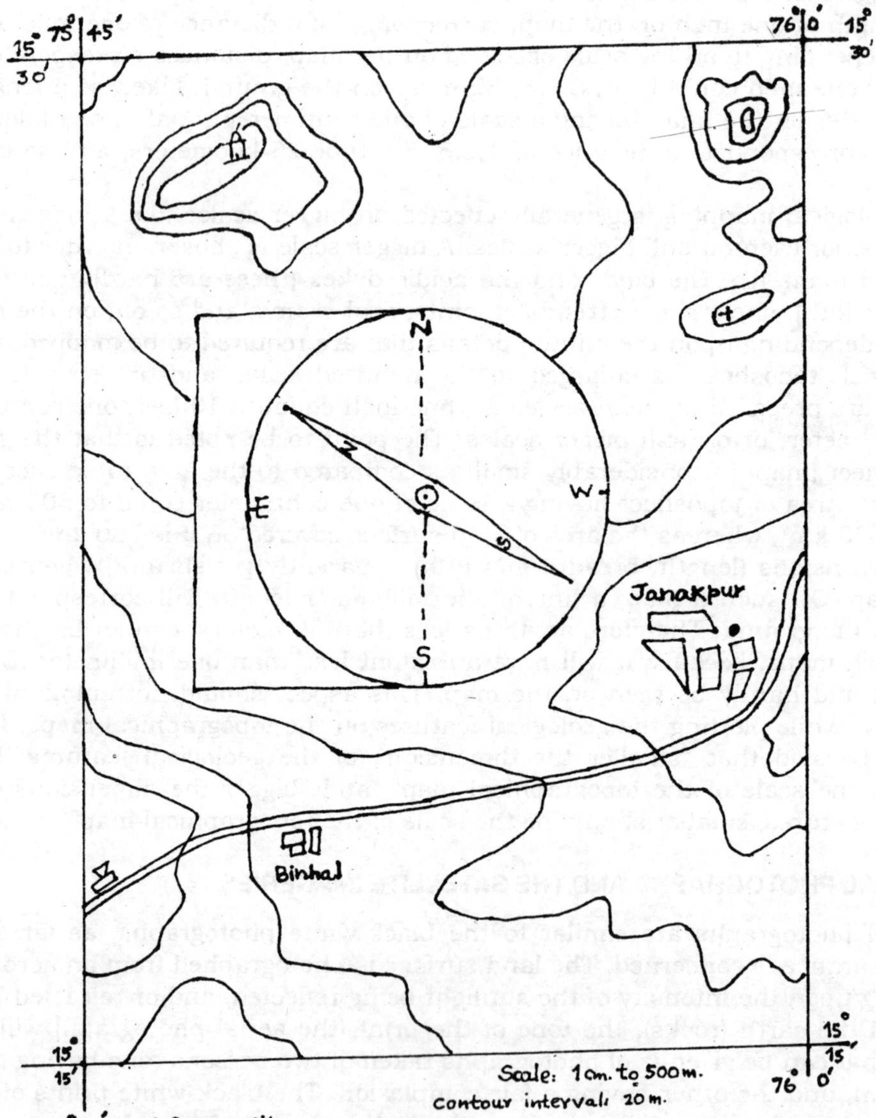

Fig. 1a. Orienting toposheet in the field using a compass. The topographical sheet (shortened expression is a "toposheet") is not oriented, because the needle of the compass has, not coincided with the N-S line (dotted line) of dial of the compass. Note that the N-S line of the dial is however parallel to the N-S line (Longitude) of the toposheet. Rotate the toposheet in an anticlockwise direction (this refers specifically to the situation in Fig 1a) without disturbing the position of the compass on the toposheet, till the N-S line of the dial comes exactly below the needle of the compass. Now the toposheet is said to be oriented in the field.

or a major syncline, can be identified, because, the aerial photographs cover a vast terraine on them, unlike that contained in a topographical map. However, the petrographical and the structural features inferred from the aerial photographs, are required to be confirmed through the routine field mapping technique.

The satellite imageries are presented in coloured and black-white prints. These cover considerably extensive area, as compared to that accomodated on the aerial photographs. The method of deciphering the rocks and the structures, is from the colour difference (if the print be coloured one), or from the tonal variation (when the print is in black-white). Whatever deciphering is made, it is required to be confirmed through the routine field mapping technique.

THE SECOND COMPONENT

This is a heterogenous component, as it embraces magnetic instrument, lens, chemicals, and so on. Each one has an important role to play, in collecting data in the field, and is therefore indispensable. In the following paragraphs a detailed account of the said components, has been presented.

CLINOMETER COMPASS

This instrument plays a dual role. The first use of it is to take the bearings in the field, in order to locate oneself on the map, during the field work. The second use is to measure the amount of dip, the direction of dip, the trend of shears, and the other directional structural elements. A documentation prepared on a paper, with the marking of north line, and a scale, is technically described as a "Map". The topographical sheets, the geological maps, or any other similar map, therefore bear a north line, as well as a scale. The features like the "hills, the valleys, a temple on hill top, the streams, and so on", are shown exactly in a way, as they are found to exist with respect to each other on the ground. The geographical positions of the different objects are marked with respect to the NORTH LINE drawn on the map. The distance between the various objects, are monitored by the scale chosen for the map.

The first step during the mapping exercise, is to "orient the topographical map". The meaning of this expression is, that the NORTH LINE as marked on the map, should be made parallel to the geographical (actual) north direction. This is achieved through the use of the clinometer compass. The compass is placed over the topographical map, and the line joining the north and the south ends, as marked on the dial of the compass (this line is exactly parallel to the bridge of the clinometer compass), is made parallel to the NORTH LINE drawn on the map. Now the map, along with the compass placed over it, is rotated, till the magnetic needle of the compass, also coincides with the "north-south line of the dial" (Figs. 1 a,b). This entire operation is technically called as "orienting the topographical map in the field". It is of paramount importance to remember that the topographical map (so also the geological map) should faithfully and accurately document the features as they are found to occur on the surface of the earth.

It is therefore necessary to measure the directions accurately. The clinometer compass serves this purpose, besides it can be used to measure the inclination of the planar surfaces. It is therefore felt necessary to note the essential features of this instrument. A clinometer compass consists of a magnetic needle placed on a

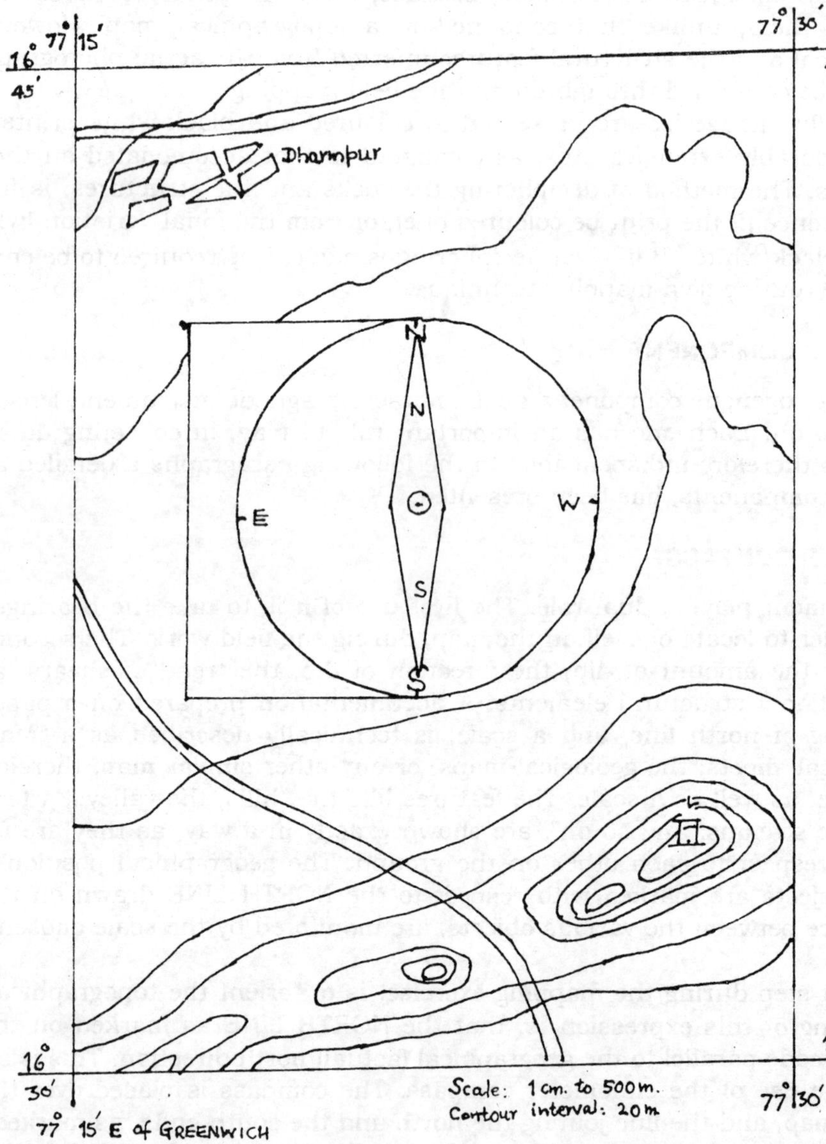

Fig. 1b. Oriented toposheet in the field using a compass. The toposheet is oriented, because the needle of the compass, the N-S line of the dial, and the N-S line (longitude) of the toposheet, are parallel to one another. This toposheet is altogether different from that shown in Fig. 1a.

pivot, such that it can move freely, when the outer case of the compass, is held in a horizontal position. The horizontality of the base of the clinometer compass, is judged through the free movement of the magnetic needle over the pivot. In the case of a Brunton compass, its horizontality is ensured by bringing the "air bubble" at the centre of the "circular spirit level". In Figs. 2 a,b and photos 1 a, b, c, a clinometer compass (this has a bridge), and a Brunton compass (this has a mirror fixed on the underside of the lid that covers the instrument), have been shown, in order to get an idea about the most important tool of the geologists.

Photo 1a: Front view of a Brunton compass. Note the three parts of the instrument-mirror (this on the underside of the lid), the Dial and the pointer. On the dial, the magnetic needle, and the index for measuring inclination, can be seen. Note that on the index, there are two "spirit levels"-one is circular, the other is elongated in shape. The circular spirit level (it is in fact a "bull's eye) is used to ascertain the horizontality of the compass, while the elongated spirit level is used to measure the angle of inclination. The "reversed position" of E and W, on the dial should be decidedly taken note of.

Photo 1b: Backside of a Brunton compass. Note the "lever" on the back surface of the "dial" part of the compass. The pointer is partially seen on the left hand side of the "dial" part of the compass.

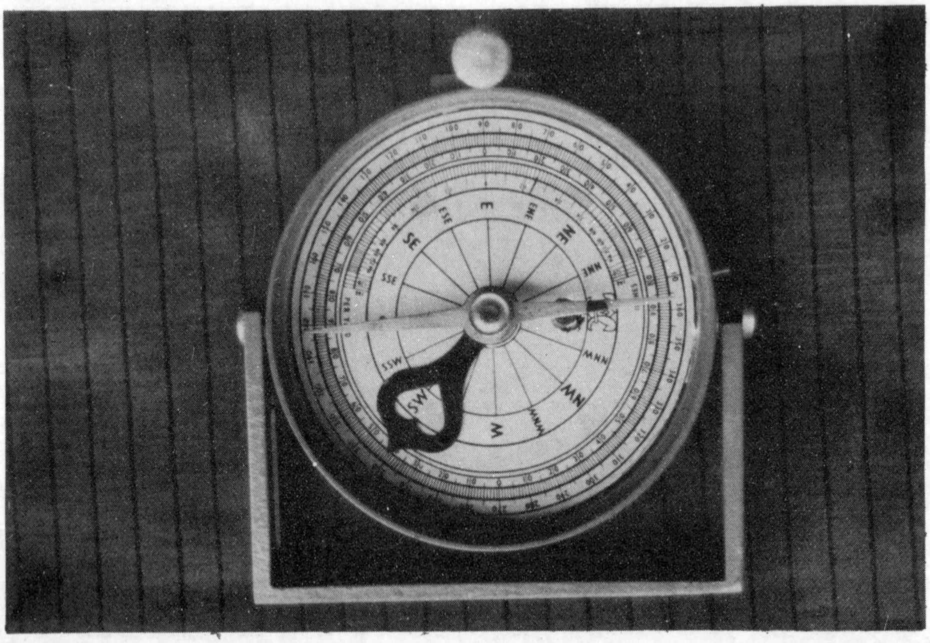

Photo 1c: Front view of Clinometer compass. Note that the bridge is fixed at the N and S ends of the dial. The dial has both the "circular and the quandrantal" graduations on it. Note that there is a "pendulum" fixed below the magnetic needle. The "big screw" towards the "east" marking, is used to arrest the movement of the needle. The "reversed position" of E and W on the dial, should be decidedly taken note of.

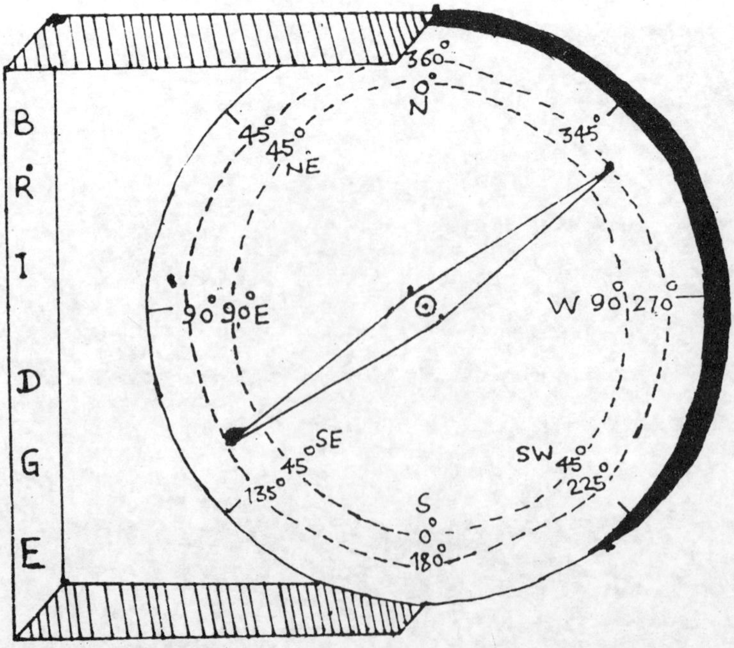

Fig. 2a. Clinometer Compass. The dial bears the Quadrantal (inner dashed circle) as well as the circular (outer dashed circle)" angular readings". The dial is practically at the base of the instrument, and beneath the magnetic needle. Also note that this compass has a "Bridge" which is fixed at the "Nand S" end of the instrument.

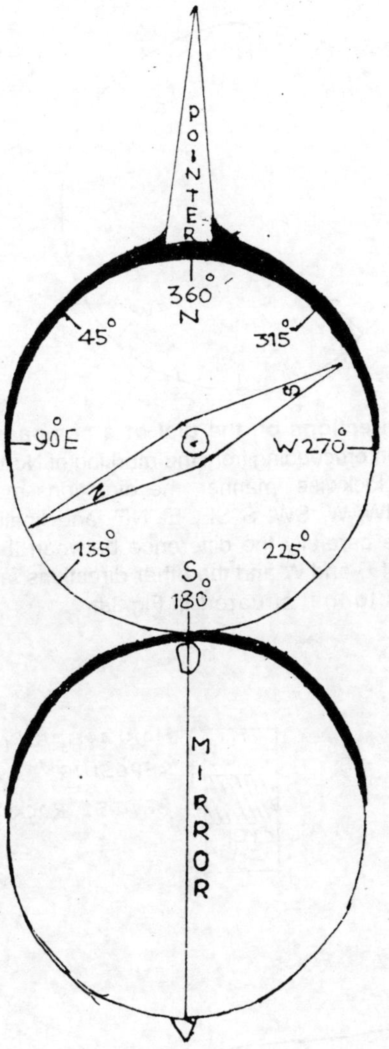

Fig. 2b. A Brunton Compass. Note that the dial bears the Circular readings only. The dial is practically at the base of the instrument, and below the magnetic needle. Also note that the instrument has a Mirror and a Pointer.

Proceeding from the north (on a toposheet or a geological map) point in a clockwise manner, the directions encountered are East, South, West, and then finally returning to the north direction (Fig. 3a). But on the dial of the compass, the positions of East and West, are reversed i.e., on the right hand side of the N-S line, west is marked, while on the left hand side, east is shown (Fig. 3b).

The positions of the east and west, are reversed on the dial of the compass, in order to get the correct reading of the direction being measured by the instrument. How it yields the correct reading, it has been elaborately described by Gokhale (1991).

The magnetic needle of the compass, covers an angular distance of 360 degrees. The directions (technically called as the bearings) are recorded in two fashions namely, a) the Quandrantal and b) the circular. In the former category, the angle of 360 is split into four quandrants, namely, the northeast, the southeast, the northwest, and the southwest, which are abbreviated as NE, SE, NW and SW, respec-

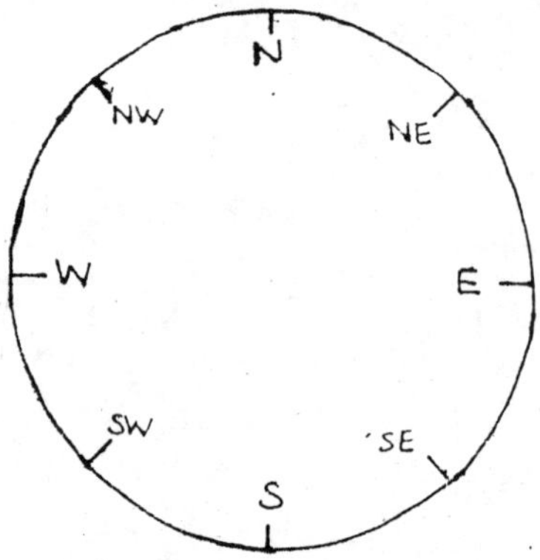

Fig. 3a. Geographic directions. Note that on proceeding from the marking of North (N) in a "Clockwise" manner, the directions met with are NE, E, SE, S SW, W, NW, and finally north. Note carefully the difference between the positions of E and W and the other directions, as compared to that situated in Fig. 3b.

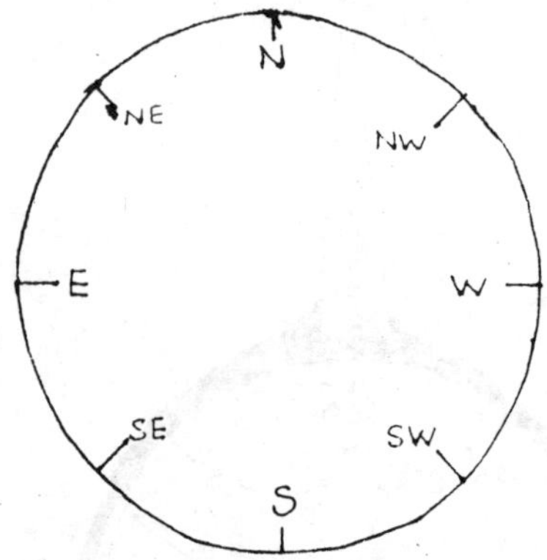

Fig. 3b. Directions on the dial of a compass. Note that on proceeding from the marking of North (N) in a "clockwise" manner, the directions met with are NW, W, SW, S SE, E, NE, and finally north. Note carefully the difference between the positions of E and W and the other directions, as compared to that situated in Fig.3a.

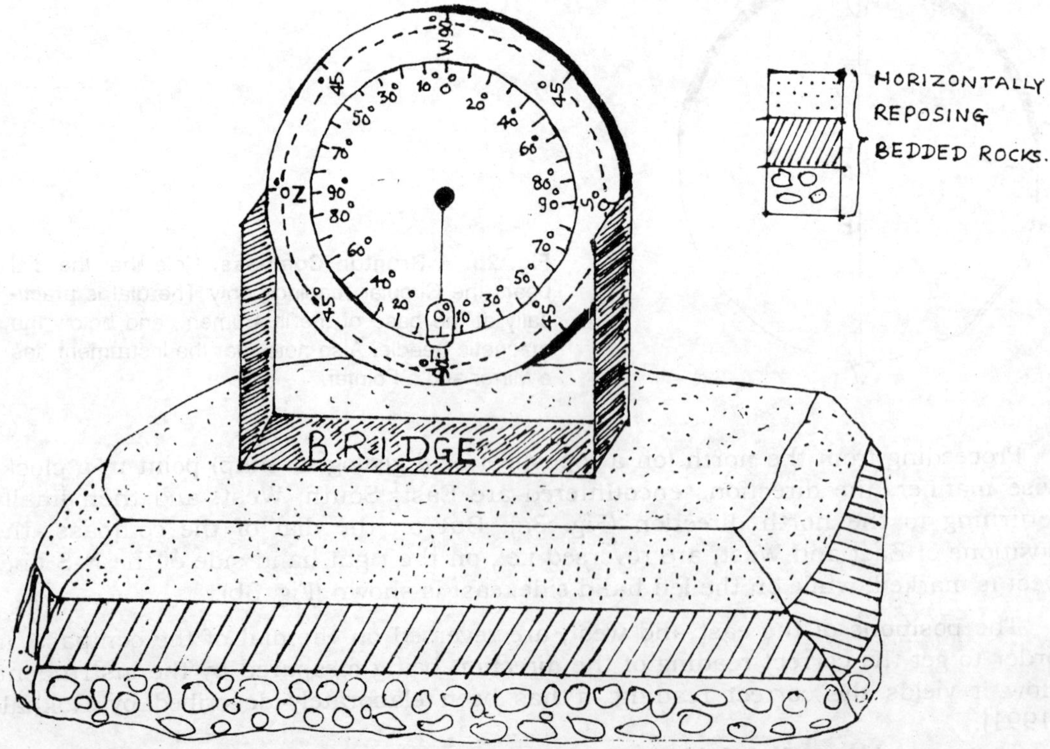

Fig. 4a. Measuring Inclination of a planar surface with a clinometer compass. Bridge of the clinometer compass could be placed on the rock, because the surface is smooth and reasonably plane. Note that the "pendulum" stands exactly against the zero graduation. The beds (rocks) are therefore horizontal (flat) in attitude. It is important to note that the position of the pendulum against the "Inner graduated circle" is to be considered while measuring the inclination of any planar surface.

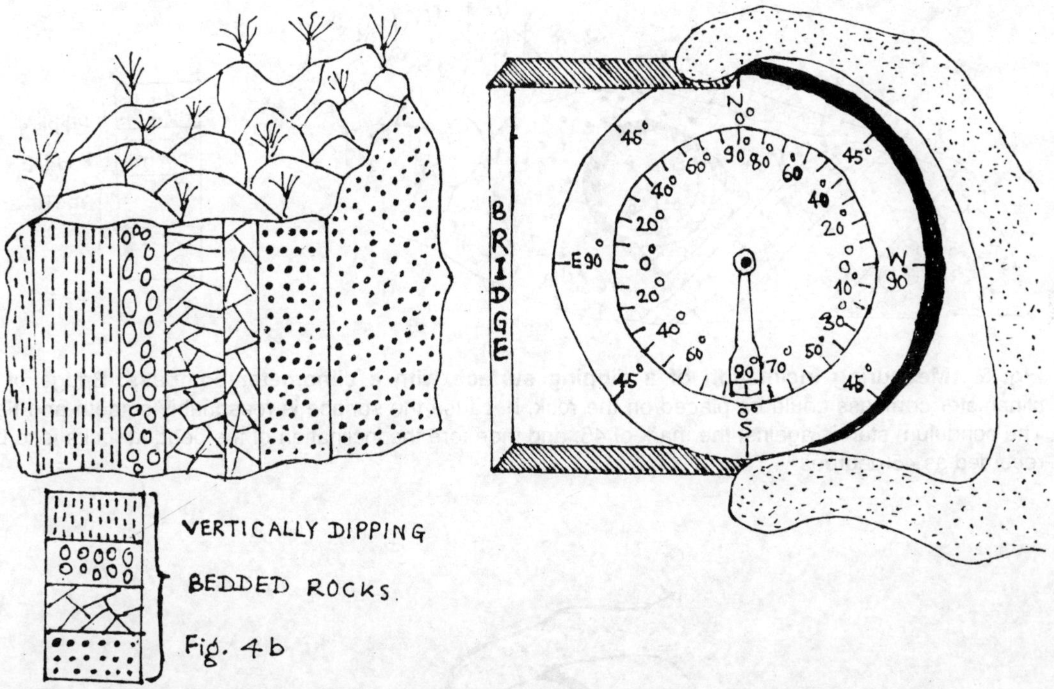

Fig. 4b. Measuring inclination of vertical surface with a clinometer compass. Bridge of the clinometer compass is held parallel to the bedded rocks, in order to measure the inclination of the bedding planes. Note that the pendulum is exactly against the graduation mark of 90 degrees. The inclination of bedded rocks is therefore recorded as vertical. Note that the position of the pendulum on the "Inner graduate circle" is to be considered while noting the inclination of the beds or any other planar surface.

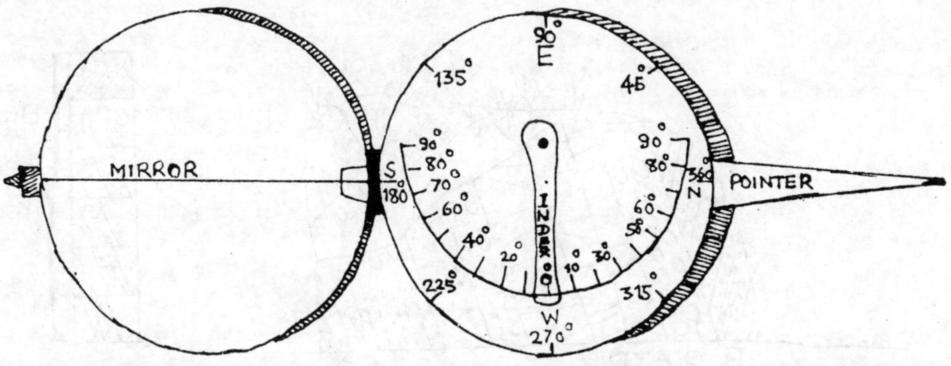

Fig. 4c. A Brunton compass. Graduations on the Outer Circle stand for the Circular readings. The arc of circle is used for measuring the inclination of planar surfaces. Unlike the pendulum of a clinometer compass, an Index is fixed on the dial. Fulcrum of the index is fixed in between the pivot of the magnetic needle (not shown in the figure), and the marking of 90 on the dial. Two "spirit levels" are attached (not shown in the figure) to the index. While taking reading of any planar surface, the index is required to be operated by the lever which is fixed at the backside of the base of the Brunton compass (Photo 1b). In Fig. 4c, the index is against zero mark of the graduated arc of circle. Unless the lever is operated, the position of the index on the arc of circle remains unchanged, even though the Brunton compass is held in horizontal, vertical or any other position.

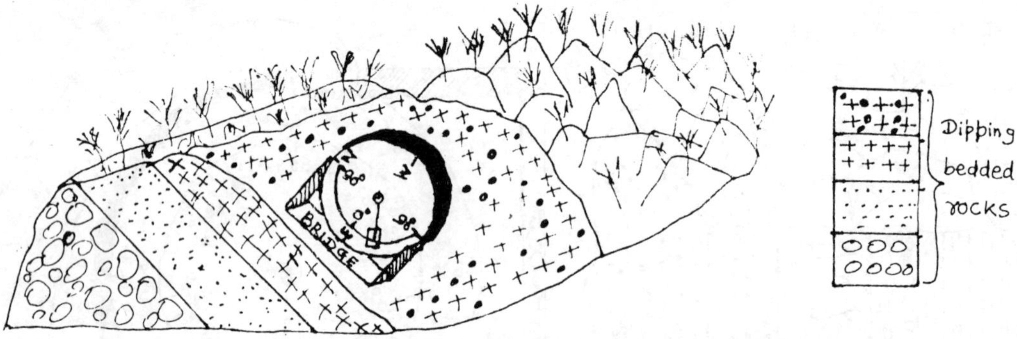

Fig. 5. Measuring inclination of a dipping surface with a clinometer compass. Bridge of a clinometer compass could be placed on the rock, because the surface is reasonably smooth and flat. The pendulum stands against the mark of 45, and therefore the inclination of the bedded rocks is to be recorded as 45 degrees.

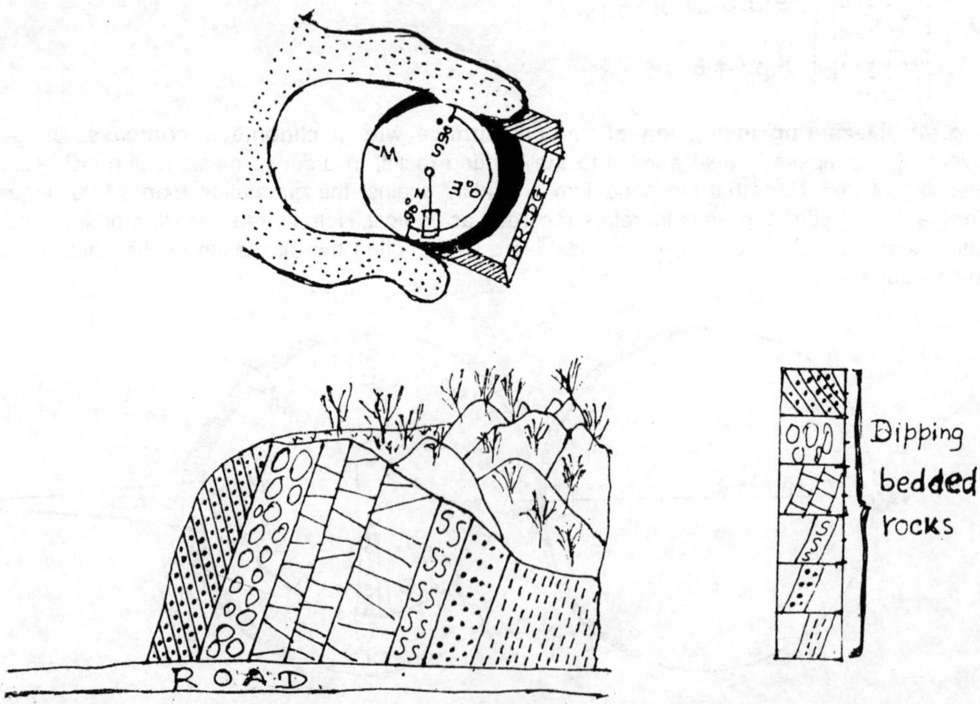

Fig. 6. Measuring inclination of dipping bedded rocks with a clinometer compass. Bridge of the clinometer compass could not be placed on the surface of the bedded rocks, because it is neither smooth nor plane in attitude. However the bedding planes are distinctly observable in the road cutting. Therefore the bridge is held in the air so that it is parallel to the bedding planes. The position of the pendulum against the graduated circle should be immediately read off, without disturbing the position the instrument. In Fig.6, the pendulum stands against the value of 60 degrees. Therefore the inclination of the bedded rocks is to be recorded as 60 degrees.

tively. In each quadrant, the readings are marked from zero to 90 degrees. The N and the S markings on the dial, stand for zero degrees, while the E and the W markings stand for 90 degrees (Fig. 2a).

In the circular bearings, the marking of N on the dial, stands for zero or 360 degrees. The reading of 90 stands for the east direction, 180 for the south direction, while 270 stands for the west direction. The values of the angle, increase in an anti-clockwise manner (Fig. 2b). At this juncture, it is felt necessary to explain asto why the compass is called as a "clinometer compass". Unlike "the land surveyors', the geologists have to record the inclination of the "bedding planes of rocks, the joint planes, the fault planes, the plunge of lineation, and so on". This aspect can not be achieved by the magnetic needle of the "Surveyors compass". The word "clino" means angle (inclination), and "meter" means a measurer. Therefore a "clinometer compass" is an instrument that measures the inclination of the planar surfaces from zero (horizontal) to 90 degrees (vertical), besides being useful in measuring the directions.

In order to measure the inclination of the planar surfaces, a "non-magnetic strip" is fixed below the needle of the compass (in the case of a clinometer compass), and it is called as a "pendulum", because it swings like the pendulum of a clock. In the case of a Brunton compass, the strip is fixed between the pivot of the needle, and the "E" marking on the dial. These features have been shown in Figs. 4a, b,c and those can be seen in Photos 1a, c.

In the case of a clinometer compass, the "pendulum" rests on the dial, when the instrument is monitored to take the bearings, trends of shears, and so on. In such a position of the compass, the pendulum is ineffective, as far as the measurement of the inclination of a planar surface is concerned. When the dial is made vertical (Figs. 4a and 5), the pendulum becomes operative, and it can then register the inclination of a planar surface.

If the planar surface be reasonably smooth, then the bridge of the clinometer compass should be placed on the surface, whose inclination is to be measured (Fig. 4a). If the surface be not smooth, or be inclined one, then the bridge should be monitored parallel to such a surface (Fig. 4b and 6, see frontispiece photo, wherein the geologist is holding bridge of the clinometer compass parallel to the dipping beds).

The mechanism of measuring the inclination of a planar surface with a Brunton compass is a bit complicated operation. The "index" which replaces the "Pendulum" of the clinometer compass (Fig. 4c.), is required to be manipulated by the "lever" which is fixed at the backside of the Brunton compass (Photo 1b). The lever is manipulated till the "air bubble" of the spirit level attached to the index, comes to the centre. The position of the index on the dial, registers the inclination of the planar surface. All this manipulation is required to be picked up and mastered, through a rigorous practice and patience.

PROCEDURE OF MEASURING THE TREND OF LINEAR ELEMENTS

Whenever the direction of any linear element is required to be measured (it may be the trend of a shear, the trend of a fault), *the north end of the dial should be oriented towards the object to be cited.* This is achieved by placing the clinometer compass on the palm of the hand, and then bending the bridge on the left hand side. In fact, the bridge bends only on the left hand side of the clinometer compass. By so doing, the

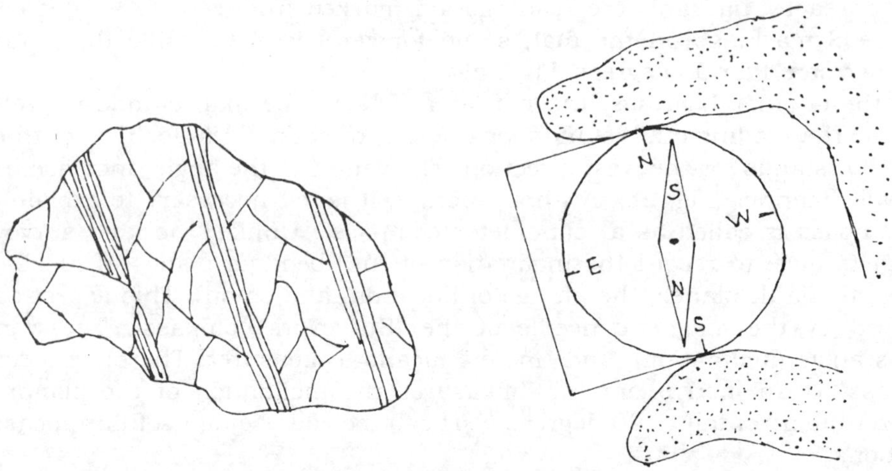

Fig. 7a. Orientation of a clinometer compass while taking trend of shears developed in the outcrop of rock. Note that the bridge of the compass is held parallel to the trend of the shears. The "N" end of the needle is read off. It gave a reading of S 20°E. Since it is a trend, the reading is recorded as N20°W-S20°E.

Fig. 7b. Orientation of a clinometer compass while taking trend of quartz veins exposed in the outcrop of rocks. Note that the bridge of the compass is held parallel to the length of the veins. The "N" end of the needle gave a reading of N60°E. Since it is a trend, the reading is recorded as N60°E-S60°W.

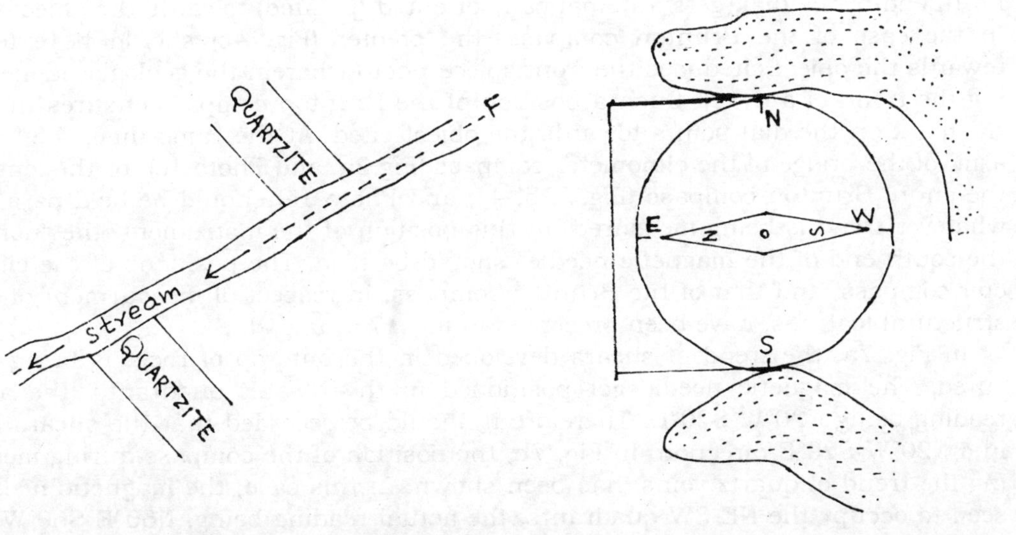

Fig. 7C. Orientation of clinometer compass while taking trend of a fault noticed in the outcrop of rocks. Note that the bridge is held parallel to the trend of the stream, which also marks the existence of fault in the rocks. The needle is observed to be exactly coinciding with the E-W line of the dial of the compass. Therefore the trend of the fault is also recorded as E-W.

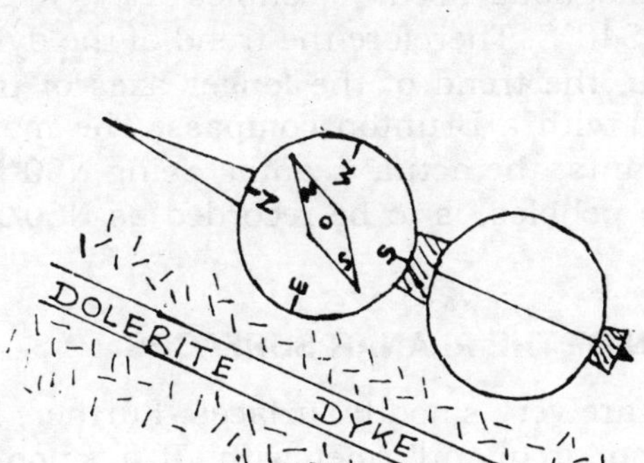

Fig. 7d. Orientation of a Brunton compass while taking the trend of a dolerite dyke traversing rocks. Note that the entire length of the compass is held parallel to the length of the dyke. The "N" end of the needle gave a reading of N40°W. Since it is a trend, the reading is recorded as N40° W-S40°E.

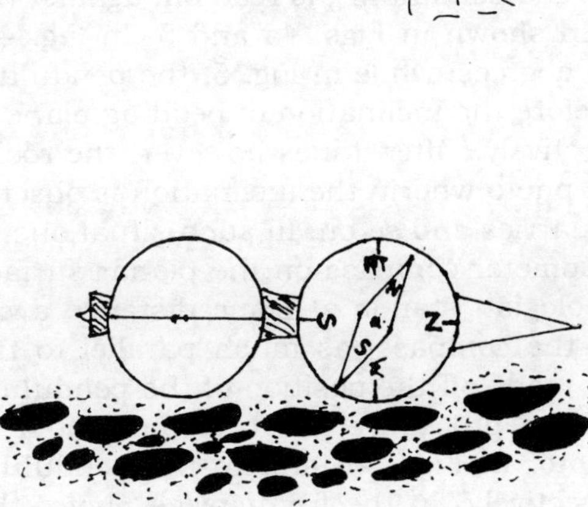

Fig. 7e. Orientation of a Brunton compass while taking direction of elongated pebbles of a conglomerate. Note that the entire length of the compass is held parallel to the longer axes of the pebbles. The "N" end of the needle gave a reading of N60°E. Since it is a trend, the reading is recorded as N60°E-S60°W.

north end of the dial, gets automatically oriented (pointed) towards the object cited. In the case of the Brunton compass, the pointer (Fig. 4c) should be extended towards the object cited (see the frontispiece photo wherein the geologist is measuring the trend of a dyke). Such a position of the Brunton compass ensures that the "N" mark on the dial points towards the object cited. At the same time, the "longer arm" of the bridge of the clinometer compass (Fig.2a, and Photo 1c), or the length of the entire Brunton compass (Figs. 2b, 4c, and Photo 1a) should be held parallel to whatever trend is being measured. In this position of the instrument, the "north or the south end of the magnetic needle" should be read. The positions of the clinometer compass, and that of the Brunton compass, in respect of measurment of some structural features, have been presented in Figs. 7 a, b, c, d, e.

In Fig. 7a, the trend of shears developed in the outcrop of rock, is being measured. The magnetic needle got positioned in the NW-SE quadrants, the actual reading being N20°W-S20°E. Therefore it should be recorded that the shears trend in a N20°W-S20°E direction. In Fig. 7b, the position of the compass during measuring the trend of quartz veins, has been shown. In this case, the magnetic needle is seen to occupy the NE-SW quadrants, the actual reading being, N60°E-S60°W. The trend of the quartz veins is to be recorded as N60°E-S60°W. In Fig. 7c, the position of the compass while taking the trend of a fault, is shown. In this case, the magnetic needle is oriented exactly along the E-W direction. The trend of the fault is therefore to be recorded as E-W. In Fig. 7d, the positioning of the Brunton compass, while taking the trend of a dyke, is shown. The magnetic needle occupies the NW-SE quandrants, the actual reading being, N40°W-S40°E. Therefore the trend of the dyke is to be recorded as N40°W-S40°E. In Fig. 7e, the trend of the longer axes of the pebbles of a conglomerate, is being measured with a Brunton compass. The magnetic needle is oriented in the NE-SW quandrants, the actual reading being N60°E-S60°W. Therefore the trend of the axes of the pebbles, is to be recorded as N60°E-S60°W.

PROCEDURE OF MEASURING THE INCLINATION OF THE PLANAR SURFACES

The bedding planes of the sedimentary rocks, are very smooth surfaces. Further, in the field, the outcrops of such planes, are not frequently met with. If a smooth surface be available, then the bridge of the clinometer compass should be placed over it, and the position of the "index mark of the pendulum", is read off, against the graduated dial of the compass. This has been shown in Figs. 4a and 5. In Fig. 4a, the pendulum stands against a value of zero degrees, while in Fig. 5, the pendulum stands against the value of 45 degrees. Therefore the inclination of bedding plane is to be recorded as zero and 45 degrees, respectively. Often times however, the rocks get exposed in road cuttings (see frontispiece photo wherin the inclination is observable in a road cutting), railway cuttings, in quarries and so on. In such situations, it is not possible to place the bridge of the clinometer compass on the planar surface (bedding planes). Therefore the observer (geologist) stands at some distance away from the exposure, and holds the bridge of the compass, as much parallel to the dipping planar surface as possible, and then reads off the position of the pendulum against the graduated dial of the compass, The procedure has been presented in Figs. 4b, and 6, and in the frontispiece photo. In Figs. 4b and 6, the pendulum stands at values of 90 and 60 degrees, respectively. In the frontispiece photo, the pendulum stands at a value of 70 degrees.(bottom side of photo).

The procedure described above, is to be extended to other planar surfaces, like the joint planes, the fault planes, the planes of schistocity and so on. Like the clinometer compass, the Brunton compass too, can be used to measure the inclination of the planar surfaces. As already noted, the pendulum is replaced by an index, and the latter is required to be monitored by a lever. Whereas a pendulum of the clinometer compass swings freely, the index of the Brunton compass is required to be manipulated by the lever, wherever inclination is to be measured. While manipulating the lever, the spirit level attached to the "index" should be watched, and the "air bubble' should be brought into the centre of the "longer spirit level" (Photo 1a for the spirit levels, 1b for the lever). It is said as the "longer spirit level" because by the side of it, is a circular spirit level. This latter spirit level is in fact a "Bull's eye", and it is utilised to ensure the horizontality of the compass, while taking the readings of directions. During the entire operation of measuring the inclination of any planar surface, the case of the Brunton compass should be held parallel to the planar surface. The position of the index on the graduated arc on the dial (Fig. 4c) need not be read off in that position of the compass, *because, even if the position of the compass be changed, the index does not move from its position, as long as the lever is not operated. In other words it means that the "reading of inclination", gets locked up, i.e., it does not change.* This is an advantage over the clinometer compass, because the position of the pendulum on the dial is required to be read off, without disturbing the position of the instrument. The procedure of taking the readings of the trend and the inclination of the planar surface has been described and illustrated with field photos by Gokhale (1991). Whether it is a clinometer compass, or a Brunton compass, only through rigorous practice, these instruments can be handled properly.

THE ROLE OF A POCKET LENS

This usually has a magnification of 10 (ten), and it helps to identify the minerals encountered in the rocks, megascopically i.e., in a hand specimen. Such an identification is fairly accurate. Later on, the identification made in the field, is to be confirmed by the microscopic examination of the thin sections prepared from the rock specimens. At least the commonly and the frequently encountered rock forming minerals, should be identified, while, conducting the field mapping exercise. Unless this is achieved., the distinction between the igneous, the sedimentary and the metamorphic rocks, can not be made during the field studies. Even though the terrain be made up of say, by only one category of rocks (either igneous, sedimentary of metamorphic), there exist many varieties in it. Therefore sample of a rock collected from an outcrop, has to be broadly identified. Unless a change in the mineral composition is warranted, a fresh sample of rock exposed in a nearby place, need not be collected. As already noted, the purpose of collecting the samples of rocks, is to get an idea (knowledge) about the kind or kinds of rocks constituting the terraine that has been undertaken for the field mapping programme. The number of samples should neither be insufficient, nor too many. It is therefore necessary to identify the minerals in the hand specimens, and to broadly identify the rock samples. This procedure avoids duplication of samples. The proper way of using the pocket lens in identifying minerals in hand specimen, has been shown in Photos 2a and 2b.

Photo 2a. The geologist (Dr. G. S. Pujar) is holding the pocket lens far away from his eye, in order to observe a magnified view of the specimen. This however does not help to study the features like the cleavages, inclusions, transparency of the grain, and so on. This is an incorrect way of using the pocket lens.

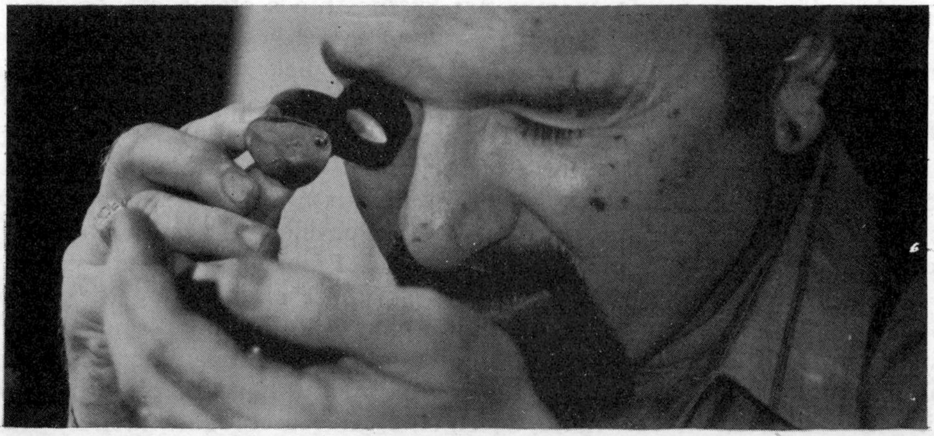

Photo 2b. Note that the pocket lens is held closest to the eye (like viewing through the eye-piece of a petrological microscope), which gives a view as though one is looking through a petrological microscope. The focussing is effected by moving the specimen, up and down, but not the lens. This is the correct way of using the pocket lens.

It is necesary to be familiar with certain specific, megascopic characters that are useful in identifying minerals composing the rocks, that are collected during the field mapping exercise. Huang (1962) has given a detailed account of the megascopic characters of the commonly encountered minerals in the igneous rocks. These have been considered here with partial changes. The frequently occurring minerals in the igneous rocks are quartz, potash felspar, plagioclase felspar, pyroxenes, amphiboles, and mica. The characteristic metamorphic minerals are garnet, staurolite, kyanite, talc, steatite, asbestos, actinolite, wollastonite, corundum. In the case of the sedimentary rocks, not many minerals are met with. Further, only one or two of them compose the entire composition of the rock. As already noted, the sedimentary rocks are almost monomineralic in composition, and therefore, the distinctions are made more on the grain size and their shapes, than on the mineral composition. The diagnostic properties of the commonly encountered minerals, are given below. It is to be remembered that, such of the properties that are observable in hand specimen, during the field work, like the hardness, the streak, the habit, the cleavage, the colour, have been considered.

QUARTZ AND FELSPARS

These two minerals are present in all the three varieties of rocks, and hence it is necessary to identify them, and make a distinction between them. These two minerals are colourless, as compared to the ferromagnesian minerals (pyroxenes, amphiboles, micas). Therefore one could be mistaken for another. The distinctive property is the absence of cleavage in quartz. Therefore, the colourless grains encountered in the specimen of rock, should be carefully examined to ascertain asto whether the grains possess cleavage planes. Its absence indicates the grains to be of quartz, while its presence indicates the grains to be of felspar. A further distinction of felspars into potash felspar and plagioclase felspar, is necessary. In this context, the observations of Huang (op.cit) are noteworthy. He writes "The distinction between orthoclase and plagioclase is best accomplished by observing, with lens in bright light, the presence of minute parallel lines (twinning striations) on certain cleavage faces in the plagioclase, or their absence in the orthoclase. In addition, the orthoclase is flesh coloured or reddish, and the plagioclase many times grayish (p.85)". Further, the distinction is also made on the basis of the transparency or otherwise, of the grains. The laths (grains) of the plagioclase felspars are ":water clear", i.e., these are transparent, while those of the potash felspars are turbid in appearance. This property of turbidity or transparency can be easily observed with a pocket lens. The potash felspars (orthoclase) alone show pink colour, but every lath of orthoclase, need not be expected to be pink in colour. *The point of importance is that, the plagioclase felspars, are not pink in colour.* Thus the cleavage, the colour, the turbidity, and transparency, are utilised in the identification of quartz and felspars.

FERROMAGNESIAN MINERALS

These minerals are coloured and hence the criterion of colour, cannot be employed in the identification of the different ferromagnesian minerals. Mica (biotite) is distinguished from the pyroxenes, and the amphiboles, on the basis of the shape and presence of perfect basal cleavage. Mica is flaky (as it possosses a perfect cleavage) in habit. With the help of a penknife, if the coloured grains yield flakes, the presence of mica gets confirmed. Another feature is the habit and the shape. Mica does not display prisms or equant grains. It is important to note that the mica possesses only one set of cleavage, unlike two sets that are possessed by the pyroxenes and the amphiboles. With experience, it is possible to observe, whether the grains possess one or two sets of cleavages. Therefore these features should be carefully and critically studied with the help of the pocket lens.

The distinction between a pyroxene and an amphibole, is to be made on the habit of these two groups of the ferromagnesian minerals. Between the two groups, the grains of the pyroxenes tend to be more equant (i.e., nearly equal in length and breadth), while those of the amphiboles are prismatic and tabular in form. This prismatic habit is also described by the term "accicular", which in simpler word means "needles". It might also be said that the prisms of the amphiboles are more long and slender. This feature is more clearly observable in the metamorphic rocks like the hornblende schist, the kyanite schist, the actinolite schist, and so on. The further distinction within the pyroxenes, and the amphiboles, requires an examination of the thin section, under a petrological microscope. Since this book is dealing with the field studies, the laboratory investigations have not been considered.

The metamorphic minerals are identified more on the texture displayed by the rock, but there are some typical minerals too, which could be recognised in hand specimens of rocks. The garnet is identified by the granular, equant shape of the grains, and its typical pink colour. It is also harder than the pyroxenes and the amphiboles. The kyanite displays the characteristic blue colour, and a bladed form. As already noted, the actinolite displays the accicular (needles) form. The talc and the steatite (though not frequently met with), are to be identified by their softness and a soapy feel. Calcite has a resemblance to the felspars, because both are white in colour. The calcite however shows typical rhombohedral cleavage, and it is considerably softer than the felspars.

The identification of the commonly occurring minerals in the rocks described above, helps to draw inference regarding the existence of an igneous, a sedimentary, or a metamorphic rock. A number of varieties are present in the igneous rocks, as compared to that in the sedimentary, and the metamorphic rocks, There are five families, namely, the acidic, the intermediate the basic, the ultrabasic, and the unsaturated, within the igneous rocks. Within each family, there are 4-5 distinct sub-varieties. Therefore in the field, the geologist should be able to make a broader identification of the rock-family. A scheme has been prepared to enable the field geologist, to identify the rocks in the field, from the minerals identified, megascopically from the hand specimen. It is presented below.

(a) Presence of quartz and potash felspars, indicates an acidic rock like a *granite.*

(b) Presence of biotite mica, besides quartz and potash felspars, practically confirms the rock to be a *granite.*

(c) Dominance of potash felspars together with amphiboles (generally hornblende) as the ferromagnesian mineral, indicates an intermediate igneous rock, like a *syenite.* Such a rock does not normally contain quartz.

(d) Dominance of plagioclase felspars together with pyroxenes as the ferromagnesian minerals, indicates basic igneous rocks like *diorite, gabbro.* These rocks also do not contain quartz or potash felspars.

(e) Extremely dark colour, predominance of pyroxenes, heavy in weight, indicates ultrabasic rocks like the *pyroxenite, the peridotite, the dunite, the picrite.*

However, these rocks require a microscopic examination of the thin sections of rock.

(f) Total absence of quartz raises the possibility of unsaturated igneous rocks like *the leucite syenite, the nepheline syenite* and so on. These rocks too, require a microscopic examination of thin sections, in order to identify the said unsaturated minerals.

(g) Extremely coarse grains of quartz and very big laths of felspars, thick sheafs of biotite, indicate the rock to be a *pegmatite.* This rock typically occurs as dykes and veins, traversing granites.

(h) Presence of quartz and felspars, displaying a saccharoidal texture, and a total absence of any ferromagnesian mineral, indicates an *aplite.* This rock is extremely light in colour, and it also occurs as dykes and veins, traversing the granites.

(i) Presence of only plagioclase felspars, indicates an *anorthosite.* This is a monomineralic rock, and it is found to be associated with gabbros.

The above noted combination of minerals, and the other features described above, will help *to broadly identify the igneous rocks during the field mapping exercise.*

The identification of the sedimentary rocks, is considerably simple, because these rocks are mostly monomineralic in composition. Further, the size and the shapes of the grains are required to be taken into account, in the identification of the rock sample. The mode of identification of the commonly encountered rocks, is given below.

(a) Pebbles, cobbles, boulders, set in a matrix - a *conglomerate.*

(b) Angular and irregularly shaped pieces of rocks set in a matrix - a *breccia.*

(c) Rounded to subrounded grains of quartz set in a matrix- a *sandstone.*

(d) Extremely fine grained clay particles, set in a matrix - a *shale,* a *mudstone.*

(e) Rock giving copious effervescence with dil. HCl acid - a *limestone.*

(f) Rock not reacting to dil. HCl acid, but resembles a limestone - a *dolomite.*

(g) Pebbles and boulders are too large, and are set in a matrix - a *boulder bed.*

(h) The pieces are rather angular in shape, but these are smaller in size than those encountered in a conglomerate, and are set in a matrix - a *grit.*

(i) Reddish brown colour and a porous structure - a *laterite.*

(j) Light brown to grayish white in colour and a porous structure.
It is not as hard as a laterite, when it is hammered - a *bauxite.*

(k) A rock resembling a sandstone, except possessing a gray colour. It has some resemblance to a breccia too. The rock is a *graywacke.*

The identification of the metamorphic rocks, is once again carried out by a combination of texture and typical metamorphic minerals. Another salient feature is the absence of any matrix for the mineral grains (which is always present in the case of the sedimentary rocks). The identification of the commonly encountered metamorphic rocks, is given below.

(a) Grains of quartz displaying sutured (cerrated) borders, and composing the entire rock. a *quartzite.* It is pertinent to remember that in a sandstone, the grains of quartz are rounded to subrounded in shape.

(b) Granular calcite possessing sutured borders, compose the entire rock - a *marble.* The rock reacts to dil. HCl acid.

(c) Rock composed of quarts, felspars and mica, and displaying foliation - a *granite gneiss.* This rock is light gray in colour.

(d) A greenish black coloured rock displaying perfect cleavage - a *schist.* There are two subdivisions within the schists.

(i) A fine grained rock yielding very thin slabs, when it is hammered - a *slate.* The cleavage displayed by this rock is called as the "slaty cleavage".

(ii) Depending upon the dominant ferromagnesian mineral, varieties of schists are identified, like *the biotite schist, the chlorite schist, the hornblende schist,* and so on.

It is very important to note that between a gneiss and a schist, the latter rock is more dark in colour, and it possesses better cleavage. A gneiss displays foliation.

(e) A greenish coloured rock containing metamorphic minerals, and displaying a granular texture, unlike the schistose or the gneissic textures - a *granulite.* Numerous varieties are recognised within the granulites on the basis of the dominant ferromagnesian mineral like a *pyroxene granulite, a hornblende* granulite and so on.

THE OTHER COMPONENTS

A penknife is useful in ascertaining whether the ferromagnesian mineral is a mica. Ofcourse, this is possible, only if, the rock be reasonably coarse grained one. Likewise, the hardness of a mineral could be tested, especially to distinguish a grain of quartz from that of a felspar, or that of a calcite from a grain of felspar. For this purpose too, the rock should be coarse grained one, in order to scratch the grain with a penknife. The streak of the coloured minerals, can be obtained with a penknife. Example is cited of the micaceous hematite, which is also called as the specularite. This mineral being flaky in habit, it might be mistaken for a biotite mica. The bladed form (prismatic habit) also could be mistaken to that of the prroxenes or the amphiboles. However on scratching such a mineral with a penknife, if "cherry red" streak be obtained, the identification changes to a micaceous hematite, and not to a biotite mica, or to any ferromagnesian mineral (pyroxene, amphibole). However, the use of a penknife requires skill, practice and experience.

Dil. HCl acid is very useful during the programme of field mapping. It is difficult and dangerous to carry it during the actual collection of the rock samples, and the process of mapping in the field. Therefore the suspected carbonate mineral or a rock, can be tested with the dil. HCl acid, after returning to the camp. The carbonate minerals or the rocks are confirmed by treatment with the dil. HCl acid. Calcite gives effervescence, but aragonite (its composition also is CaCO3), does not show any reaction. The role of the dil. HCl acid is especially useful when the matrix of the sedimentary rocks, is to be ascertained during the mapping exercise. If the rock gives effervescence, carbonate minerals (calcite), and not siliceous or argillaceous minerals form the matrix for the grains of the rock.

Sample bags are especially required, if fragile material or fossils, are to be collected. *Camera* is a very, useful tool, as it faithfully documents the field setting of the rocks, special features like the ripples marks, the current bedding, pillow structure, pahoehoe structure, intersecting intrusions (dykes and veins), striations/grooves on the fault planes, and so on. Seeing is believing, and therefore, the field photos many times provide a critical and an indisputable data, which can be later considered as an evidence. For every field photo taken, the place of the photograph, the direction of the view of the camera, and other information must be immediately noted, right at the site where the photograph is taken.

CHAPTER 3

Components of Geological Map

INTRODUCTION

A geological map is made up of two components; one is the rocks exposed in the area (the petrographical component), and the second is the structural features (the structural component), developed in those rocks. Usually both are documented together on one and the same map. However, if considerable variation in rock types, as well as numerous structural features be present, it is advisable (and it is better) to prepare two separate documents (maps); one showing the different rock types alone, the other showing the various structural features possessed by the rocks. This will avoid complexity and confusion, because, the boundaries of the rock types are shown by lines, and the different structural features (these are all linear elements like the faults, the joints, the folds, the shears and so on) are also required to be shown by lines. However, the petrographical and the structural data are collected together and not separately, during the process of the field work and mapping.

THE PETROGRAPHICAL (ROCK) COMPONENT

In the crust of the earth, the igneous, the sedimentary, and the metamorphic rocks are met with, but it is not necessary that all the three or two categories, must be encountered in each and every area taken up for the mapping programme. Therefore, before setting out for mapping, it is better to have some idea about the mode of occurrence of the three varieties of rocks. The igneous rocks alone occur as intrusive or extrusive bodies. The extrusive forms comprise of the volcanic erruptions, and or, as thick or thin lava flows. The volcanic eruptions often produce conical hills. Therefore the presence of such hills in any area, warrants the presence of the volcanic rocks. The volcanic erruptions at times give rise to craters. Morphologically, these give rise to large sized depressions on the surface of the earth, which might later give rise to the lakes. The famous LONAR lake of Maharashtra, is an outstanding example. Hence while studying the topographical maps, the aerial photographs, or the satelite imageries, of the area to be taken up later for field mapping, if a depression of a large size be noticed, it might turn out to be the site of a volcanic crater. Hence one could reasonably expect volcanic rocks at the site of the crater. The lava flows are often horizontal in attitude, and the resultant topography, is one of the flat topped hills. Thus while studying the topographical (contour) maps, the aerial photographs, or the satelite imageries, a "Mesa or Butte" topography be noticed, it could as well be due to the flat lying lava flows. The Deccan Traps practically covering the entire area of Maharashtra state, India, is a classic example of such a morphological scenery. The Pavagarh hills of Gujrat state; India, also display flat topped hills, and these hills are composed of flat lying flows of rhyolites.

The intrusive igneous rocks constitute batholiths, stocks, bosses, dykes and sills, besides other forms. Out of these forms, the laccoliths and the dykes produce characteristic geomorphical features. Laccolith has a domical form, and it gives rise to the structural feature called as "qua versal dip". In actuality however, the igneous rock composing the laccolith, does not possess any dip, (in fact no igneous instrusion on its won, can give rise to dipping surfaces), but the bedded sedimentary rocks into which the igneous rock (magma) had intruded, the sedimentary rocks on the top of the laccolith, produce a "hat-shaped structure". As the beds of the sedimentary rocks, dip away in all directions (like that in a dome, the laccolith thus acquires its domical form.

Photo 3. Typical "Tor scenery" is presented in this photograph. Note that the boulders are resting one over another, in a precarious manner. It is to be noted that the boulders are "insitu" ones, and not brought from some other place, which are then called as the "rolled ones". This scenery is also described as "watermelon topography, or as balanced rocks". Tor scenery is invariably produced in the granitic terraines (Courtesy Dr. V.B. Koppad).

The dykes, especially of basic igneous rocks, like the dolerite, the basalt, being usually quite resistant to weathering, these give rise to linear ridges. Therefore, while reading the topographical maps, the aerial photographs, and the satellite imageries, if a ridge morphology be noticed, the possibility of the presence of a dyke, is to be borne in mind. Of course, hard, resistant sedimentary rocks (like the sandstones), or the metamorphic rocks (like the quartzite, the banded hematite quartzite, banded jasper quartzite), can also give rise to the ridge morphology. The ultimate identification is to be made during the field studies (mapping exercise). The batholiths, the stocks, and the bosses, give rise to the "Tor scenery" which is also called as the "watermelon topography, Balanced rocks" (Photo 3). It is pertinent to note that the igneous rocks alone produce such a morphology. Another salient and an important aspect is that the acidic igneous rocks, namely the granites and the granodiorites, constitute the batholiths, the stocks, and the bosses. Therefore this feature is a very useful, and a reliable guide in ascertaining and building up the "Petrographic Component" of the geological map.

The sedimentary rocks are bedded formations, and therefore these invariably display the characteristic and indisputable geomorphic forms like the "Dip slope and Escarpment scenery, the Mesa and Butte hills, the Flat topped hills". These distinct geomorphic forms, go a long way in inferring the petrography (rocks) by studying the topographical maps, the aerial photographs, and the satelite imageries. The metamorphic rocks share the features of the igneous and the sedimentary rocks. Therefore, generally from the geomorphological point of view, it is not possible to distinguish the metamorphic rocks from the igneous or the sedimentary rocks. The metamorphic rocks therefore, definitely require an actual examination of rock specimens in the field, as well as in the laboratory.

The account given so far, is to assist the geologist in inferring the kind of the rocks, from a systematic and a careful study of the topographical maps, the aerial photographs, and the satelite imageries. The "Petrographical Component" can be thus partially built up, before setting out for the field mapping exercise.

THE STRUCTURAL COMPONENT

Unlike the "Petrographical Component", a great majority of the structural features can not be inferred from a mere study of the topographical maps, the aerial photographs, or the satellite imageries. Exception is that of the faults. A straight course of a river, the site of a waterfalls, acutely meandering stream or a river, a gorge, these are often times found to coincide with the faults. The igneous, the sedimentary, and the metamorphic rocks, possess certain structural features of their own. Thus, the sedimentary rocks alone possess bedding planes, and these are required to be traced during the field mapping programme. Most important aspect is recording the direction and the amount of dip (inclination) of the bedding planes. The surface of contact of the bottom-most bed of a succession of beds, is required to be traced in the field. Such a surface is called as the surface of "Unconformity". The metamorphic rocks possess planes of schistocity. The direction and the amount of dip of schistocity, are required to be recorded. The folded structures can be formed, only if the rocks be bedded ones. The folded structure is more commonly encountered in the metamorphic rocks. Many times, the anticline or the syncline is not readily and directly noticcable, because of erosion and/or the paucity of outcrops. Therefore the folded nature of the rocks, is required to be established by recording the direction of dip of the bedding planes. On plotting such a data, if the beds be found to dip away from each other, an anticline gets established. If the beds be found to dip towards one another, a syncline gets established. In case a fold be found to be clearly and completely exposed, the attitude of the axis of the fold, is required to be recorded. If the axis be found to possess inclination, the fold gets classified as a "plunging fold".

There are many structures that are found to be possessed by the three categories of rocks. These structures are, the joints, the shears, and the faults. The joints and the faults, belong to the category of the "planar structures". In this case, the direction and the amount of dip of the joint planes or the fault planes, are required to be recorded. The shears belong to the category of the "linear elements", because these give rise to the linear fractures in the rocks. The trend of the shears, is also required to be measured.

It is pertinent from the account pertaining to the "Structural Component" that it is not possible to infer the nature of the structural elements very easily, from a study of the topographical maps, the aerial photographs, and the satelite imageries. Actual field observations are absolutely necessary. The structural elements are shown on the base map ((the topographical map), by way of lines and arrows (in the case of the joint places, the bedding planes). Therefore a line drawn on the map may stand for a fault, a thin dyke, a bedding plane, a major joint plane or some other structural features. It is to be borne in mind that the field data are collected on a topographical map (called as the base map). The topographical maps themselves consist of closed and/or open contours, roads, river, streams, and so on. All these features are shown on the map, as lines. Therefore, all these lines will add to the already described complexity, when the petrography and the structural features are also marked as lines, on a common base map. Therefore it is better, that the structural features, and the petrographical features, are shown (marked) on two separate base maps. Likewise a separate map should be reserved for marking the spots (sites) of collection of samples of rocks.

SUMMARY

The two components of the geological map are the petrographical (rocks) and the structural features. Reasonably reliable information can be obtained, regarding the two components, by a careful study of the morphological features, revealed by the topographical maps, the aerial photographs, and the satellite imageries. This information should be confirmed or rectified during the field mapping exercise.

CHAPTER 4

Methods of Geological Mapping

INTRODUCTION

The preparation of a geological map from the data collected during the field work, comprises of marking the outcrops of the several rocks, as well as the several structural features possessed by them, on the base map (i.e., the topographical map). The igneous, the sedimentary, and the metamorphic rocks, possess different petrographical and structural features. Their mode of occurrence in the crust of the earth, is also different. Therefore, the methods to be adopted in mapping these three rock units, are found to be different. The methods are described in the following paragraphs.

SOME COMMON PROCEDURES

The first step is to get familiar with the topography of the area as presented in the base map (the topographical sheet). As already mentioned, certain inferences could be drawn, in respect of the petrographical and the structural features of the rocks, from the shapes of the hills, the shapes of the streams and the rivers, the shapes of the valleys, and so on. The presence of the ridge-morphology, suggests the existence of vertical beds (rocks), or a basic dyke (dolerite, basalt). A hill with a steep slope on one side (the contours will be found to be very closely spaced to one another), and a gentle one in the diametrically opposite direction (on this side, the contours will be found to be spaced farther apart from one another), is indicative of the hill being composed of gently dipping sedimentary rocks. Such a topography is described as the "Dip slope and Escarpment scenery". A very straight course of a stream or a river, acutely meandering stream or a river, indicate presence of a fault, or closely spaced shears. A gorge is invariably observed to be the site of a fault.

Therefore, the topographical map, the aerial photographs, the satelite imageries, should be critically studied, before setting out for the mapping exercise, specifically in respect of the shapes of the hills, the valleys, the streams, the rivers, and even the sea coasts. The topographical maps also have the bench marks, sites of temples, idgah, fort, and other prominent "land marks", on them. These are very useful during the process of locating the position of oneself (the geologist) on the map, corresponding to his position in the field, while carrying out the field work.

The second step is how to locate oneself on the map, during the mapping exercise. In order to achieve this goal, bearings are taken to at least to two known objects/spots, that are marked on the topographical map. Such objects could be, a temple built on a hill top, or an idgah, or any other prominent monument that has been marked on the map. These objects should also be distinctly visible to the person (geologist), from the spot (his position in the field), from where he is desiring to take bearings. It is further necessary that the two objects/spots to be cited, are quite wide apart from one another. The term "bearing" means the direction in which the object/spot is located with respect to the observer (geologist). The direction from the object/spot, back to the observer (geologist), is called as the "back bearing".. Thus, if the value of bearing be say N50° W, the value of the back bearing will be S50°E. This aspect of "bearing and the back bearing" has been shown in Fig. 8. The procedure of taking bearings to two objects, is presented in Fig. 9. There are certain guide lines to be followed, while taking the bearings to the objects. The optimum angle between the two bearings is found to be of 90 degrees, because, only then the intersection of the two back-bearings will be very sharp. It is advisable to maintain an angular spread of 80 to 110 degrees, between the two bearings. These features have been shown in Figs. 10 a, b, c.

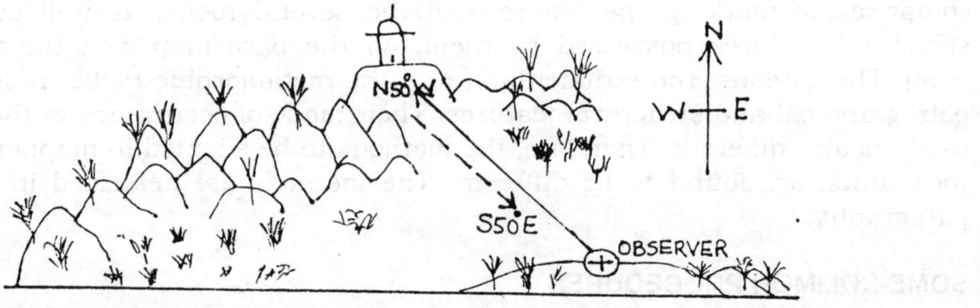

Fig. 8. Relation between bearing and back-bearing. Observer has cited a temple situated on the hill top, the reading being N50°W. This is called as "bearing or fore-bearing". Therefore when viewed from the temple, the position of the observer will be S50°E. This latter reading is called as "back-bearing".

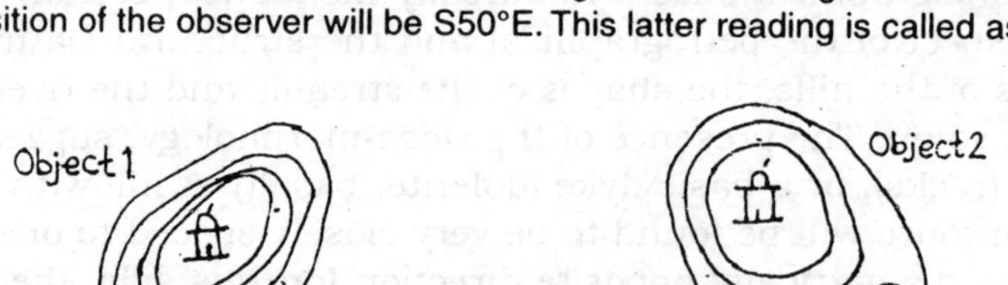

Fig. 9. Procedure of taking bearing to two objects. Locate two prominent objects that are clearly observable, like those shown in the figure. Also try to maintain a wide angle between the two readings (bearings) as has been obtained between N40°W and N55°E, (an angle of 103 degrees in this figure).

After taking the bearings to two objects/spots from the point of observation (the position of the geologist in the field), rays (lines) are drawn from the two objects (these are marked on the topographical map by the surveyor), in the reverse direction to that of the bearings. These two rays are made to intersect. This point of intersection, marks the position (location) of the observer (the geologist) on the map. This procedure has been presented in Fig. 11. It is suggested that the subsequent points of observations in the field, be located with the help of the previous one, in order to save time. Therefore on the second occasion it is not necessary to take bearings to two objects, only one is enough. The procedure to be adopted is to

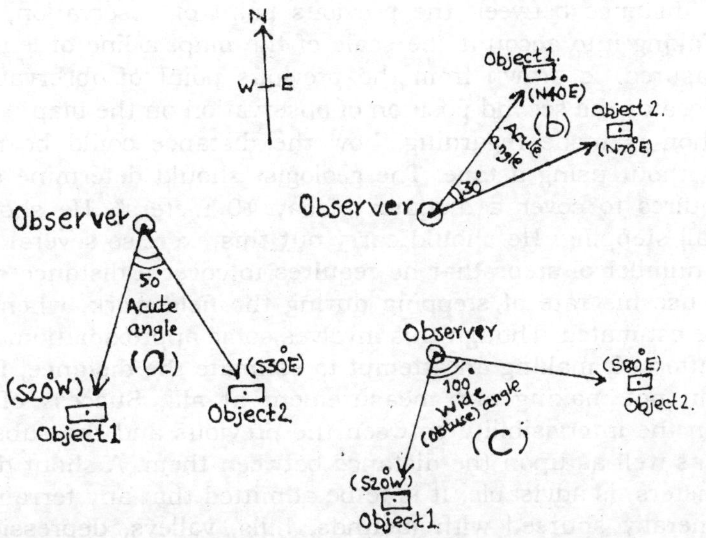

Fig. 10a, b, c. Ideal angle between two bearings. In Fig. 10 a and b, angle between the two bearings, is acute (less than 90°). With such two bearings, when the back-bearings are plotted, the point of intersection will not be sharp. This causes inaccuracy in locating position of the observer on the map. In Fig. 10c, the angle between the two bearings is quite wide (more than 90°). This ensures a sharp point of intersection after plotting the two back-bearings. This will, in turn, effect accurate location of the observer on the topo sheet (the base map). It is advisable to maintain a difference of 90°, between the two bearings, during the process of locating the position of oneself (the observer) on the base map.

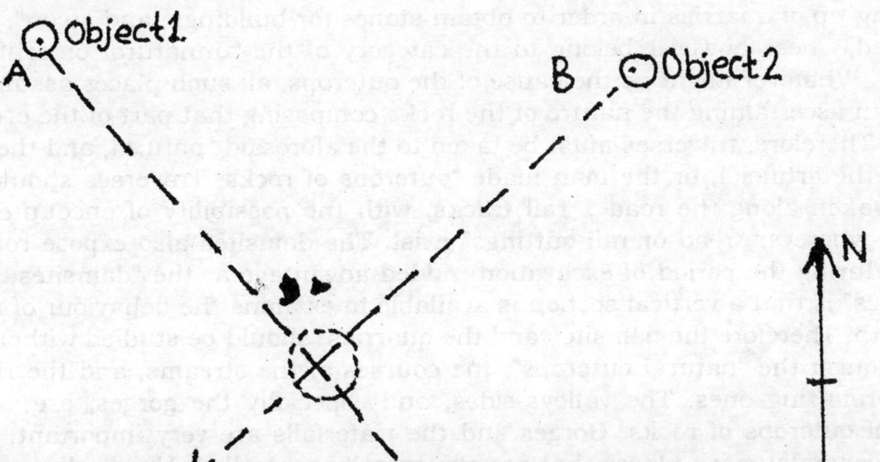

Fig. 11. Procedure of locating observer's position on base map. Intersection of lines AA_1 and BB_1 marks (locates) the position of the observer on the base map (the topographical map). Note that the point of intersection is quite sharp, because the angle between the two bearings is wide (90 degrees).

measure the distance between the previous point of observation, and the subsequent one. Taking into account the scale of the map, a line of length equal to the distance measured, is drawn from the previous point of observation. The end of such a line, locates the second position of observation on the map.

A suggestion is made regarding, how the distance could be measured, fairly accurately, without using a tape. The geologist should determine asto "how many steps he requires to cover a distance of say 10 meters". He should keep to his normal (usual) stepping. He should carry out this exercise several times, and then find out the number of steps that he requires to cover a distance of 10 meters. He should then use his rate of stepping during the field work, whenever distance is required to be estimated. Though this involves some approximation, yet it is not very inaccurate. After all, making an attempt to estimate the distance, fairly accurately, is better than not making any measurement at all!. Success of this procedure depends upon the intervisibility between the previous and the subsequent point of observation, as well as upon the distance between them. A shout distance of say a few tens of meters, is advisable. It is to be admitted that any terrain is seldom very flat. It is generally spursed with mounds, hills, valleys, depressions and so on. Inspite of it, maximum precaution and care is required to be taken to achieve a reasonably good accuracy in locating the several positions of observations made during the field work, on the base map (i.e., the topographical map).

WHERE TO LOOK FOR OUTCROPS

As already mentioned, the portion of the rocks exposed at the surface of the earth, is called as the "outcrop" of the rocks. Generally, hard and resistant rocks, occur as "natural outcrops", the softer ones quickly get weathered and become soil. It is said as the "natural outcrops" because, the human activities, also produce outcrops of rocks. Thus, "in the construction of the railway tracks, the National Highways, opening up of quarries in order to obtain stones for buildings, and so on", rocks get exposed. These however belong to the category of the "unnatural or artificial outcrops". Whatever might be the cause of the outcrops, all such places assume importance in ascertaining the nature of the rocks composing that part of the crust of the earth. Therefore, traverses must be taken to the aforesaid "natural, and the unnatural or the artificial, or the man made "outcrops of rocks. Traverses should also be undertaken along the road / rail tracks, with the possibility of encountering outcrops, wherever "road or rail cuttings" exist. The damsites also expose rocks especially during the period of excavation. Added advantage at the "damsites, and rock quarries" is that a vertical section is available to examine the behaviour of the rocks at depth. Therefore the damsites and the quarries, should be studied without fail.

Amongst the "natural outcrops", the courses of the streams, and the rivers, are very promising ones. The valleys-sides, and especially the gorges, are vulnerable sites of outcrops of rocks. Gorges and the waterfalls are very important, because, these generally mark places that are structurally controlled. Unusually meandering courses of streams and rivers, also are vulnerable sites of natural outcrops of rocks.

The hard fact and the truth is that, irrespective of the likely or the vulnerable places of encountering exposures of rocks, "every square centimeter, and every nook and corner" of the area taken up for mapping, has got to be traversed on foot. Sometimes the traverse might turn out to be a "dead one", because, outcrops of rocks might not at all be met with. Keeping such a possibility in mind, the "likely spots of outcrops of rocks" have been described above.

THE RECONNAISSANCE SURVEY

It is customary to select some area for the geological mapping. The land surface is continuous, in the context of it being limitless. Therefore certain area is chalked out for mapping during the field season. As such, before sitting out for field work (mapping), the geological literature available pertaining to the region (area) concerned should be studied. This gives considerable information, but one should not get influenced or biassed by it. Otherwise, the geological history of the area selected for mapping, cannot be rectified and/or strengthened. It is possible that some outcrops of rocks might have been missed by the previous workers. It is also possible that new quarries might have been opened up, or old ones might have been recently widened and deepened. Therefore, the new data might totally change and/or strengthen the geological map of the concerned region. The purpose of studying the previous literature, is to plan the course of mapping, with the intention of collecting data (of course, if it be existing) so far not reported from the region, verification of the previous data, or strengthening the existing geological information about the concerned (selected for mapping) region.

The reconnaissance survey is carried out to get a generalised idea about the terraine, the broad petrographical and the structural features, and the communication facilities available in the area. Rapid traverses are conducted within the area chosen for mapping. Samples of rocks might be collected, or only chips of rocks broken at the sites of the exposures, and a preliminary identification of the chips (rocks) be made. Such an identification of the chips (rocks) is to be later verified and confirmed, or it is to be rectified through a detailed and a systematic mapping, during the subsequent visit to the outcrops (exposures). The outcrops in the quarries are likewise studied, to get a preliminary "overview" of the petrographical and the structural features noticeable in the quarries. This procedure is adopted throughout the entire area undertaken for mapping. It is possible that the earlier workers might have reported some unusual petrographical and/or structural features from the area. Such spots should be specially visited during the reconnaissance survey. There is no specific period of time prescribed for conducting the reconnaissance survey. However, depending upon the "size of the area, and the nature of the terraine (rugged, forested, poor communication facilities), as well as the total period of time earmarked for the field mapping", a couple of weeks should be spent on the reconnaissance survey.

Having now equipped to undertake mapping, let us now turn our attention to the procedures to be adopted for mapping the igneous, the sedimentary, and the metamorphic rocks.

METHODOLOGY FOR IGNEOUS ROCKS

The procedure to be adopted while mapping these rocks, should suite their inherent petrographical and the structural features. Emphasis is necessary to be more on the petrography, than on the structures, because there are innumerable igneous rocks, as compared to those belonging to the sedimentary, or the metamorphic categories. However, there are some structures that are exclusively possessed by the igneous rocks. These rocks occur as intrusive or as extrusives (flows). The shape and the size of the instrusions, is generally very irregular, except that of the dykes and the sills. The distribution (occurrence) of the different varieties of the igneous rocks in the crust of the earth, is also irregular. Therefore it is necessary to collect samples of these rocks at "close intervals", and to identify the samples in the field, at least to realise that the mineralogical composition, has not changed.

The intrusive nature of the igneous rocks can be ascertained from their contact relation with the country rocks. An igneous body could be in contact with a sedimentary, a metamorphic, or even with another igneous rock. The basic (dolerite, basalt) and the acidic (aplite, pegmatite) dykes do not pose any problem, in respect of their intrusive relation with the country rocks. But with bigger igneous bodies (like the stocks, the bosses, the laccoliths, the batholiths), their intrusive character is required to be proved. If the intruded rocks (which are technically called as the country rocks) be sedimentary, or metamorphic ones, then there arise two possibilities. If the igneous body be intrusive, it is then held to be "younger" than the country rocks. This is one situation. However, if the intrusive character of body be absent (wanting), then the igneous body is held to be "older" than the country rocks. This marks the second possibility.

The problem becomes more acute, when two igneous rocks (bodies) be in contact with each other, because, which one has intruded into the other, can not be easily determined. Therefore, the contact of the igneous body with another igneous rock, should be examined very, very carefully, from a textural point of view. It is argued that the igneous rock, while it is in the magmatic stage, it is extremely hot, and in a molten condition. Naturally, when such a material comes in contact with the country rocks (which are relatively very cold), a "chilling effect" is produced. The outcome of such a chilling, is the development of a glassy or a very fine grained rock, along the border of the magmatic intrusion, with the country rocks. Therefore, an igneous rock displaying "chilled contact with the adjacent rock", is accepted to be younger, and to have intruded into the adjacent rocks. On the other hand, absence of such a chilled border, is interpreted to be older, and therefore, the igneous rock not to have intruded into the adjacent rocks. *This argument is to be considered whenever two igneous rocks (bodies) are in contact with each other.* Therefore, mapping the contact of the igneous rocks with the country rocks, is of paramount importance.

It is necessary to at least make a distinction between the acidic and the intermediate, and that between the basic and ultrabasic rocks, during the field mapping programme. Unfortunately, a sharp surface of contact between the acidic and the intermediate, or that between the basic and the ultrabasic rocks, is not encountered, within the igneous rocks. On the contrary, the igneous rocks display a gradational transition from the acidic to intermediate, intermediate to basic, or basic to ultrabasic varieties. As a contrast, the situation in the sedimentary rocks, is worth taking into consideration. The surfaces of contact between a sandstone and a shale, or that between a sandstone and a limestone, or that between a shale and a limestone, is easily distinguishable in the field. The mapping of the sedimentary rocks poses no difficulty. But the mapping of the igneous rocks is quite laborious, and it requires utmost patience, and a very keen, and acute observation of the rocks in their natural habitat (outcrops).

The geologists are forced to be content with the collection of the surficial samples of rocks. However, the existence of quarries helps to collect samples from different depths. The igneous body is a three dimensional entity. Therefore the contents and

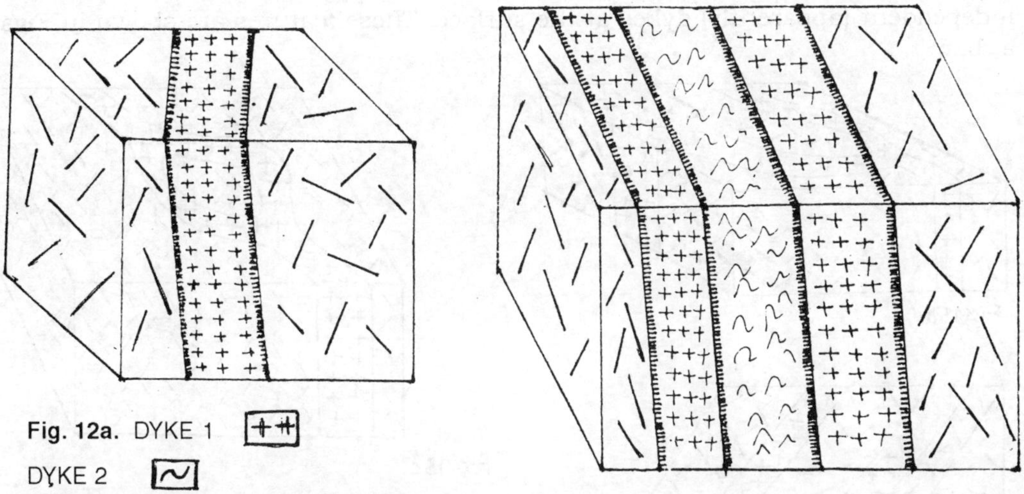

Fig. 12a. DYKE 1

DYKE 2

CHILLED BORDER OF DYKES

COUNTRY ROCKS

Fig. 12. Role played by chilled borders in deciphering number of basic dykes. **(a) Singular dyke.** Note that there are only two chilled borders, **(b) Multiple dykes.** Note that there are four chilled borders corresponding to two dykes

the variety of minerals composing it, might vary at depth, and in turn, give rise to the different varieties of igneous rocks. The bore-hole samples are welcome, but these are normally obtained only for the economic mineral deposits (technically, these are also rocks, but because these consist of ore minerals, these are termed as the "economic mineral deposits"), in order to ascertain the mineral wealth of the area. The sinking of bore-holes for the academic pursuits, is very expensive. Therefore, for the academic studies, one has to feel contented with samples from the quarries (if present), and those from the surficial exposures (the outcrops).

In the case of dykes, not only its trend is to be recorded, but its margins with the country rocks, are to be carefully studied. The dykes generally display chilled margins (the borders). It is also possible that it could be a multiple intrusion (dyke). In such cases, whereas a singular dyke displays only two chilled borders, a "Two in One Dyke", displays four chilled borders. These features have been shown in Figs. 12 a,b.

The observation of the chilled borders alone can establish the presence of a "Two in One, or a Multiple Intrusion", and therefore, this field feature (namely, chilled borders) assumes a great importance and significance. In a terraine made up of lava flows, like the Deccan Traps of India, it becomes imperative to locate those dykes, that had brought the lava up from depth, to the surface of the earth. Such dykes are called as the "Feeder Dykes". These dykes being basaltic in composition, like the Deccan Traps, the dykes could be identified, only if, "chilled borders" be established. Thus, this situation clearly demonstrates the importance of recognising the "chilled borders", wherever dykes are encountered.

The dyke intrusion when traced along its length, this might disclose the presence of branching, and the possibility of its reunion to form a "ring dyke" like structure. The branching could be even at depth, its manifestation (appearance) being, as two independent (apparently) dykes at the surface. These features are shown in Figs. 13 a, b, c.

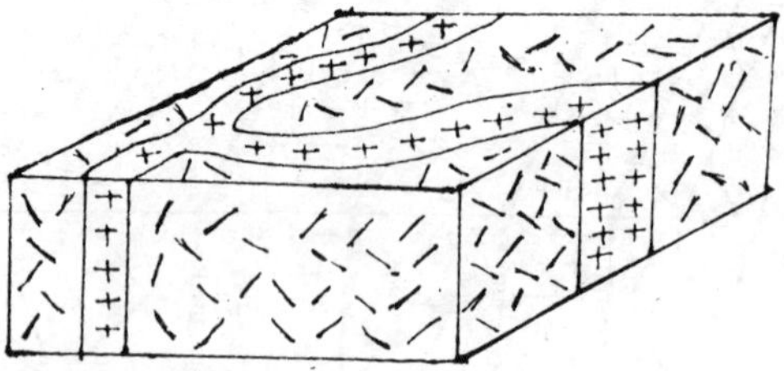

Fig13a.

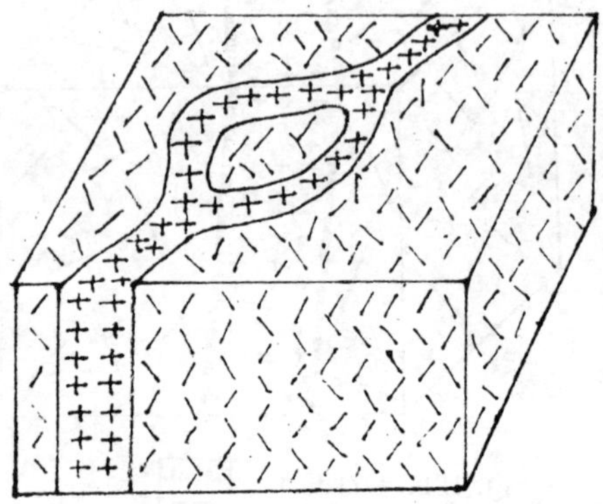

Fig.13c

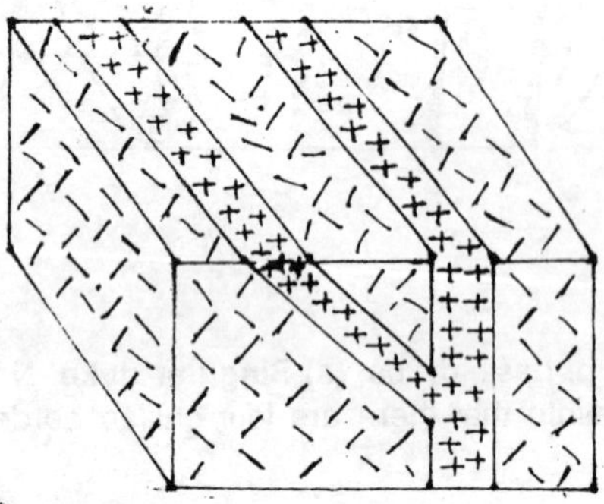

Fig.13b

Figs. 13. Branching dykes. (a) A dyke branching only at the surface. **(b)** A dyke branching only at depth. At the surface it appears to form two independent dykes. **(c)** A branching dyke producing a ring at the surface.

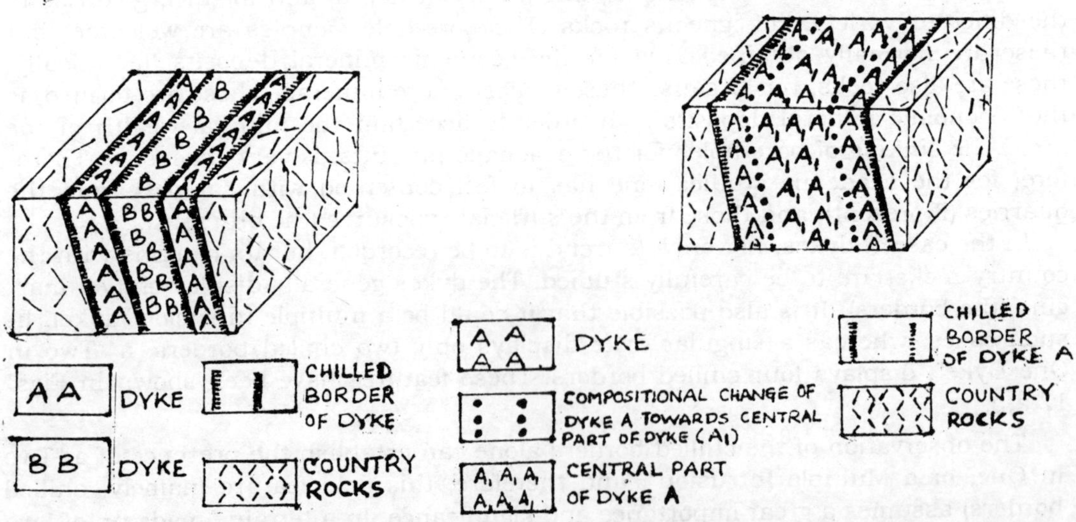

Fig. 14. Composite and differentiated intrusions. (a) **A composite intrusion**. The composition of the dykes A and B is different. Therefore these two dykes together constitute a composite intrusion. If the composition be found to be identical, then the dykes are to be categorised as "Multiple intrusion". (b) **A differentiated intrusion.** The intrusion (dyke), does not display any change of composition, on megascopic examination of rock samples collected from the border towards the interior parts. However, the subsequent laboratory studies (microscopic examination, chemical analysis) might reveal a distinct change from the marginal part towards the centre of the intrusion. It is then categorised as a "Differentiated intrusion". The point to be stressed is that serial samples must be collected from the margin towards the central part of the intrusion. Only then differentiation can be established (of course if the intrusion had really undergone such a process).

Waghàmare (1989) has reported numerous patterns of branching dykes from the northern part of the Gadag Schist Belt, Gadag district, Karnataka state. The point to be stressed is, that, it is not sufficient to merely take the trend of the dyke at some place of observation, but it should be studied all along its extent, both at the surface, and in depth (if exposed in a quarry, road cutting, and so on).

There are other varieties of dykes, namely, i) differentiated dyke, and ii) composite dyke. These dykes can not be recognised during the field studies, but only after a systematic collection of samples along the width of the dyke, and then studying the thin sections prepared from the samples, and/or from the chemical analyses of the samples, forming the course of the "laboratory studies". The point to be noted is the importance of the collection of the samples during the mapping exercise. The distinction between the differentiated and the composite intrusion (dyke), is shown in Figs. 14 a,b.

The dykes at times contain xenoliths in them. Gokhale and Koppad (1971), Koppad (1976), and Gokhale and Puranik (1981) have reported such instances from the Gadag Schist Belt, Gadag district, Karnataka state. Koppad (op. cit.) has further observed that only a part of the dyke contained xenoliths. Therefore it is suggested that whenever a dyke is met with in the field, it should be traced all along its length, till it vanishes out, with the hope of coming across xenoliths in it.

The behaviour of dykes described above, can not be shown (marked) on the base map, however bigger scale be chosen for mapping. Therefore, separate sketches are required to be drawn on convenient scales, like, 1 cm. equal to 1 metre (the scale of the base map even could be 1 cm. equal to 500 metres. On such a scale, dykes which are about a meter or two in width, can not be shown at all!. Hence separate sketches are required to be constructed).

It is rather difficult to distinguish a "sill" in the field for two reasons. In the first place, the sills are not that frequently encountered as are the dykes. Secondly, the sills are mostly composed of basic rocks, like basalt or a dolerite. Further, unless the "invaded rock" possesses planar structures, like the bedding planes, schistose planes, flow banding etc., the magma (the igneous body) can not intrude along the planar surfaces of the intruded rocks. Akin to the dykes, the upper and the lower surfaces of the sill also could develop the 'chilled borders". If the sill be sufficiently thick, it could even produce a differentiated intrusion. In fact the sills are more prone to differentiation than the dykes. Therefore, it is necessary to collect serial samples from the lower contact surface to the upper contact surface of the sill. The sills can also give rise to a "laccolithic structure". Therefore, the sills should be very carefully studied, if encountered during the field mapping exercise. The sills encountered in the sedimentary rocks, could be easily distinguished, because of the existence of a considerable colour, and a compositional (minerals) difference between them. The sills also can not be shown on the base map, because of their small thickness, like that possessed by the dykes. Therefore the sills are required to be shown as separate sketches drawn on a convenient, bigger scale, than that of the base map (the topographical map).

XENOLITHS AND INCLUSIONS

As already noted, the basic dykes at times contain xenoliths. Xenolith means a "Foreign stone". Granite which is the most extensively encountered igneous rock in the crust of the earth, it often times contains dark coloured pieces of rocks. These are called as the "xenoliths or inclusions" (Photo 4).

Photo 4. The photograph depicts "xenolithic structure". Note that the xenoliths are dark, varying in sizes, but are elongated ones. Also note that the xenoliths, in this case, are displaying parallelism between their longer dimension. Note that the handle of the hammer is placed parallel to the longer axes of the xenoliths. The xenoliths are occurring in gray granites exposed in the southeastern vicinity of Gangavati town, Raichur-Bellary districts, Karnataka state. (Courtesy- Dr. V. N. Hegde).

The disposition, the distribution, the shapes and the sizes of the xenoliths are of paramount importance in deciphering the mode of origin of the granitic rocks. Therefore, whenever the xenoliths are met with, separate and detailed sketches should be drawn, depicting the position of each and every xenolith. If these be found to possess one of their dimension greater than the others, the trend of the longer axes of the xenoliths should be invariably recorded, and accurately shown in the form sketches. The magmatic or the non-magmatic origin of the granitic rocks, is many times proved from the shape and the attitude (trend) of the longer axes of the xenoliths. Effort must also be made to identify the minerals composing the xenoliths, during the field work, because, this knowledge later helps to trace the source rock of the xenoliths. Whatever is observed in the outcrop of the xenolith, it should be faithfully and accurately documented in the form sketches. It is also advisable to take field photos, especially of the shapes and the sizes, and disposition of the xenoliths. Such photos are to be later scrutinised carefully in the laboratory. Such a study helps to classify the xenoliths into i) Cognate and ii) Accidental varieties. But this is possible, provided acute observations are made during the field studies. The xenoliths are also called as the inclusions, the schlirens, the rafts, and so on.

In conclusion it may be noted that with the igneous rocks, the mapping should be conducted on the following lines;

(a) The contact with the country rocks to be traced and studied very carefully,
(b) The disposition, the shapes and the sizes of the xenoliths to be mapped carefully,
(c) The trend of the dykes to be recorded, and these intrusions to be followed along their extent till these disappear,
(d) The samples to be collected along the width of the dykes and the sills,
(e) Separate sketches on bigger scale to be drawn for the dykes and the sills,
(f) Field photos to be especially taken, wherever xenoliths are encountered,

(g) If the xenoliths be more long than wide, the trend of the longer axes of the xenoliths to be recorded, and

(h) Numerous samples of rocks should be collected, and identified, in order to recognise varieties of rocks.

METHODOLOGY FOR THE SEDIMENTARY ROCKS

Unlike the igneous rocks, emphasis is required to be laid on mapping the structural features possessed by the sedimentary rocks, because those features are unique, and characteristic ones. Petrographical variation is very minimum, a mere five varieties being met with. These are the sandstones, the shales, the limestones, the dolomites, and the conglomerates. Mapping of the sedimentary rocks is easier than that of igneous rocks, because the former rocks possess a regular shape and size.

Further, these rocks possess bedding places which provide directions for taking traverses during the mapping exercise. It is also possible to estimate the extent of the sedimentary bodies in the crust of the earth in the three directions (two horizontal, the 3rd being the depth). Petrographically too, the sedimentary rocks contain very few minerals in them. In fact, as already noted, these rocks are categorised into different types, more on the grain size, than on the mineralogical composition. There are only four groups of sedimentary rocks, namely, a) the rudaceous, b) the arenaceous, c) the argillaceous, and d) the calcareous. However the fourth group (the calcareous) is not recognised on the basis of the grain size, but on the mineralogical composition. The calcareous group consists of the carbonates of Ca, and Mg (calcite, aragonite and dolomite). Therefore, during the field work (mapping), it suffices to estimate the grain size of the minerals, with the help of the pocket lens. The coarse, the medium, and the fine grain sizes of the constituents (the minerals) of the rock, correspond to the rudaceous, the arenaceous, and the argillaceous categories, the rock names being, the conglomerate, the sandstone and the shales (argillite, mudstone clay rock), respectively.

The sedimentary rocks however possess many primary structural features like, the current bedding, the ripple marks etc., to which a mention has been already made. *The unique feature of the sedimentary rocks, is the possibility of the presence of fossils (mega or micro) in them. In fact, neither the igneous nor the metamorphic rocks, possess fossils.*

NATURE OF MAPPING TECHNIQUE FOR THE SEDIMENTARY ROCKS

A separate sub-title is given because, the sedimentary rocks possess several primary structural features, unlike the igneous and the metamorphic rocks. Stress is therefore required to be laid more on the structural features, than on the petrographical characters. In the following paragraphs, the technique to be adopted, has been elaborated.

The traverses should be taken both in the dip, and the strike directions. The dip traverses help to map the different and all the beds constituting the sedimentary body. It is to be remembered that the sedimentary rocks are formed (laid) in a basin, and the several beds differing in grain size and mineral, are laid, one over another. Therefore it becomes necessary to establish the order of sedimentation (meaning deposition), which is technically called as the "order of superposition". In the direction of dip, progressively younger beds (rocks) are to be expected, unless those are

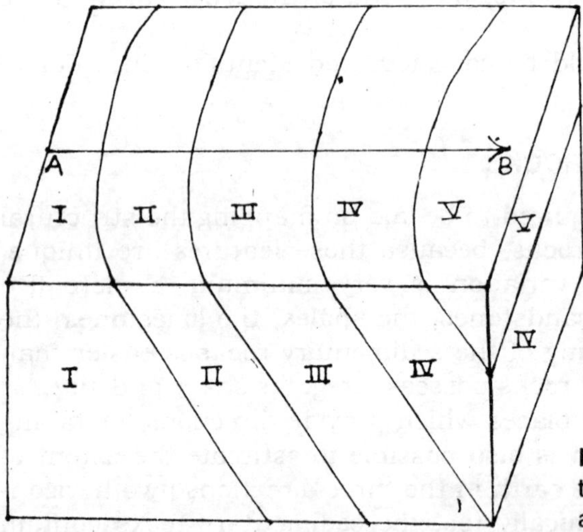

Fig. 15a. A dip traverse. Note that in only one traverse along AB (which is in the dip direction) all the beds (rocks) are encountered.

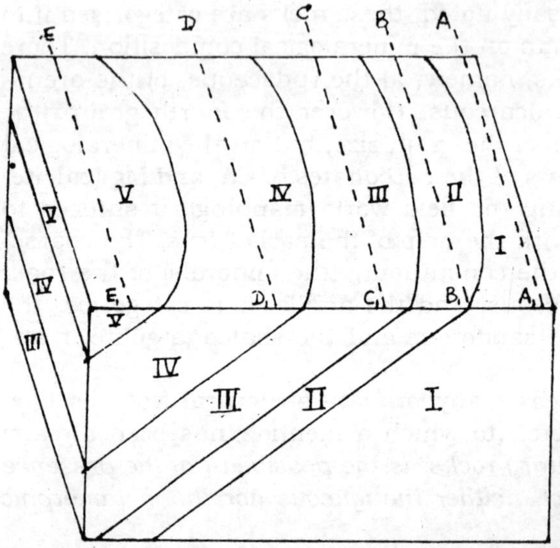

Fig. 15b. A strike traverse. Note that only one bed (rock) is encountered in each "strike traverse". Such traverses help to establish the lateral (strike) extent of the beds (rocks).

overturned during the process of folding (if folded), or during the process of the upliftment of the basin. The establishment of the order of superposition (and this is very much needed and is sought after by the stratigrapher), is possible only during the field mapping exercise. It is also possible that the direction of dip might have changed. Only on the basis of a change in the dip direction (either towards or away from some position), the folded structures can be established. Therefore the "dip traverse" is of paramount importance to the Geoscientists (stratigrapher, a structural geologist, as well a sedimentologist) because, it unfolds the entire succession of the beds (the rocks) laid in the basin.

Mapping also should be conducted in the direction of "strike" of the beds (rocks). Such a traverse helps to ascertain the "lateral extension" of the beds. However, unlike the "dip traverse", where all the beds laid in the basin, are met with, in the "strike traverse" only one bed could be traced. Therefore several "strike traverses" parallel to one another, are necessary, each traverse being confined to one bed. The features of the "dip and strike traverses" have been presented in Figs. 15 a,b.

The sedimentary rocks are laid in a basin, while the basin itself is formed in the pre-existing (and hence older) crustal rocks. The surface of contact of the floor of the basin, with the older rocks, is technically called as the "surface of unconformity". The geologists mostly talk of the relative ages of the rocks, that are encountered in the crust of the earth. It is argued that the rocks occurring in the deeper parts (levels) of the crust of the earth to be older, while those met with at the surface of the earth, to be younger or youngest in age. The concept of "unconformity" is an "indisputable tool" that has been made available to the *stratigraphers*, like the "geological thermometer" to an *economic geologist*, or the "stress ellipsoid" to a *structural geologist.*

There are several types of unconformities, but most of them pertain to the sedimentary rocks. It is argued that during one episode (period) of sedimentation, the beds (rocks) acquire certain attitude, like a dip of 40° due N20°W. All the beds laid during that episode of sedimentation, by and large, will possess the said attitude. During the succeeding episode of sedimentation, the beds (rocks) might acquire a different attitude like say, a dip of 10° due N50°W. All the beds (rocks) belonging to the second episode of sedimentation, will possess the attitude of a dip of 10° due N50°E. These two episodes of sedimentation, though enacted during the different periods, the beds (rocks) belonging to them, will be found to be in contact with each other. Apparently it gives an impression, that the several beds (rocks) were laid simultaneously.

Therefore during the field mapping exercise, the direction and the amount of dip must be recorded with meticulous care. Unless this is done, the existence of two or more episodes of sedimentation, differing in age, can not be established. The surface of contact between the beds (rocks) possessing the dip of 40° due N20°W, and those possessing the dip of 10° due N50°W, marks the plane of unconformity.

The disposition of the above described two sets of beds (rocks) belonging to two episodes of sedimentation, in the field (i.e. the outcrops), helps to assign the relative ages to them. During a traverse taken in a general northwesterly direction (i.e. in the direction of dip), if the beds with a dip due N 20°W be encountered first, and then those with a dip due N 50°W, the former set of beds are to be assigned an older age (Fig. 16a).

On the other hand, if the beds with a dip due N 50°W be encountered first, and then those with a dip due N 20° W, then the former set of beds are to be assigned an older age (Fig. 16b).

In both the situations, the traverse has been taken in the direction of dip of the beds. The point of importance to be noted is that, the "direction of dip of the beds" is utilised in establishing the two episodes of sedimentation, and assigning the relative ages to the two sets of beds.

With the sedimentary rocks, it is necessary to establish the direction in which the sediments were brought into the basin. This direction is generally derived from the position of the conglomerate horizon (if it be present, and if it be a basal conglomerate bed) in the spatial distribution of the several other beds (like the sandstone, the shale, the limestone and so on) belonging to the basin. In the absence of a basal conglomerate horizon, the position of the next "coarsest rock" namely a grit, or a coarse grained sandstone, may be considered to fix up the margin of the basin, and to derive the direction of transport of the sediments into the basin. It is argued that the sediments are predominantly brought into the basin,

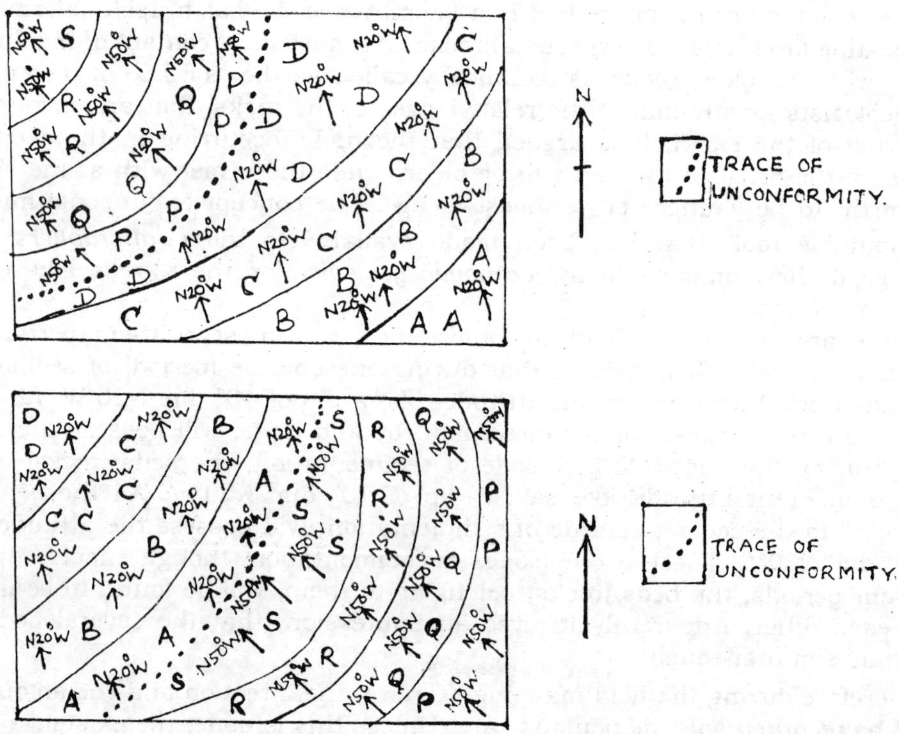

Fig. 16(a & b). Recording and plotting direction of dip of beds (Rocks). These figures elucidate the importance of plotting the direction and the amount of dip of the beds (rocks) during the mapping of the sedimentary rocks. The deposition of the beds in the area is very important and crucial while ascribing relative ages to the rocks. **(a)** Proceeding from the southeast corner of the map (area), in a northwesterly direction, it can be seen that the beds (rocks) dipping due N20°W are encountered first, and then those dipping due N50W. The former beds are therefore assigned an older age. The dotted line marks the trace of "unconformity" between the two sets of beds (rocks). The A, B, C, D set is held to be older than the P, Q, R, S set. **(b)** Proceeding from the southeast corner of the map (area) in a northwesterly direction, it can be seen that the beds (rocks) dipping due N50W are encountered first, and then those dipping due N20W. The former beds (rocks) are therefore assigned an older age. The dotted line marks the trace of "unconformity" between the two sets of beds (rocks). The P, Q, R, S set is held to be older than the A, B, C, D set.

by the rivers. As soon as the river enters the basin, its velocity is reduced, and the coarsest sediments are laid first, at or near the "margin of the basin". The sediment could be rudaceous (conglomerate) or a very coarse arenaceous grade material (sandstone).

THE ROLE OF CURRENT AND OSCILLATION RIPPLES

Whenever these primary sedimentary structures are met with, extreme caution is required to be taken to accurately record them in the form of sketches on a bigger scale (because these can not shown on the base map used for mapping, as it is small enough for the said structures). The importance lies with the fact that the "current ripples" help to derive the direction in which the sediments where brought into the basin while the "oscillation ripples" help to ascertain the "top side of the beds" (especially in such cases, where the beds are vertical, or are in an overturned position). It is advisable to take field photographs of the disposition of the said structures. Such a documentation carries more conviction, and promises acceptability amongst the geologists, the world over.

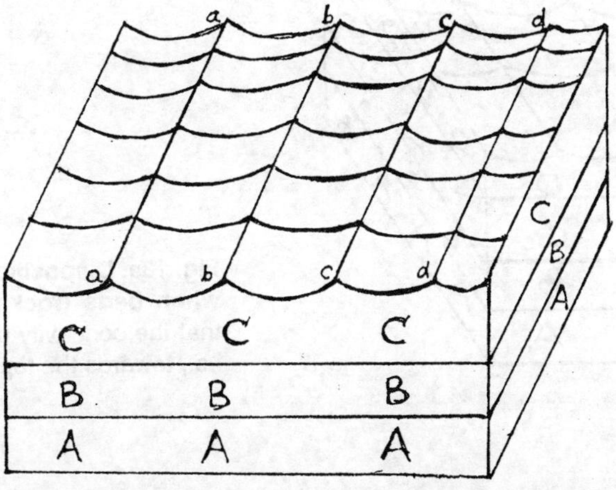

Fig. 17a. Oscillation ripples. Note the concavity and the symmetrical shape of the ripple. A, B, C are the several beds (rocks) laid one over another. C is the top most bed. Important feature to be observed is that the concave side of the ripple is pointing upwards, and hence towards the top side of the bed C. aa, bb, cc, and dd are the axes of the ripples. The direction of concavity of the oscillation ripples helps to establish the top side of the beds (rocks), especially when the beds are folded.

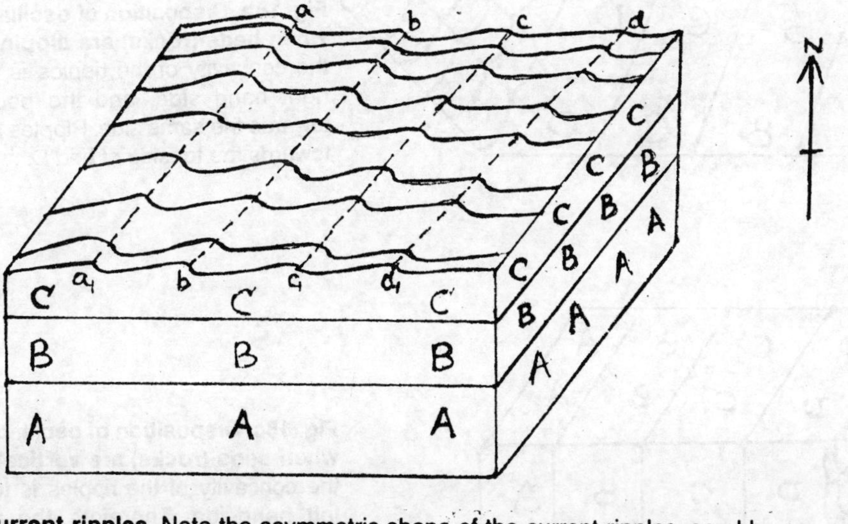

Fig. 17b. Current ripples. Note the asymmetric shape of the current ripples. aa_1, bb_1, cc_1, and dd_1 are the axes of the crests of the ripples. The axes trend in a general NE-SW direction, and the wave surface (ripple) is steeper on the easternside. The direction of movement of the current, is therefore considered to be from west to east. A, B, C are the different beds laid one over another, and the ripples are produced on the top of bed C. Note that the position of the "steeper slope" of the asymmetric ripples, holds the clue in deriving the direction of movement of the current.

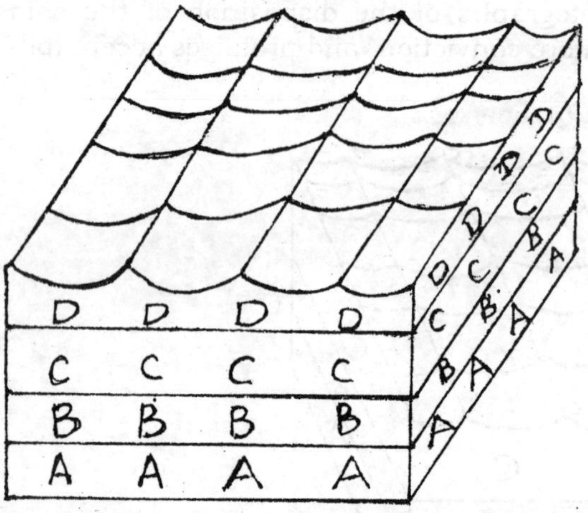

Fig. 18a. Disposition of Oscillation ripples when beds (rocks) are horizontal. Note that the concavity of the ripples is upwards i.e., towards the top of the bed D.

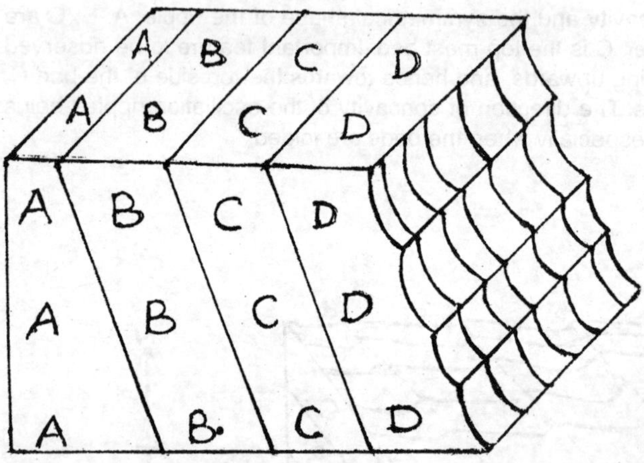

Fig. 18b. Disposition of oscillation ripples when beds (rocks) are dipping. Note that the concavity of the ripples is towards the right hand side, and the beds also dip towards the same side. Ripples are concave towards the topside of bed D.

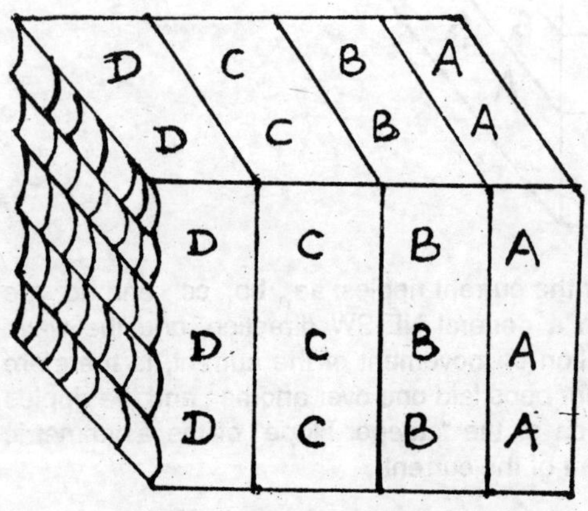

Fig. 18c. Disposition of oscillation ripples when beds (rocks) are vertical. Note that the concavity of the ripples is towards the left handside. Therefore the top of the vertical bed D is also towards the same side. **In the absence of the oscillation ripples, the top-side of the beds could not have been established : it could be on the left or the right hand side, because the beds are vertical in attitude.**

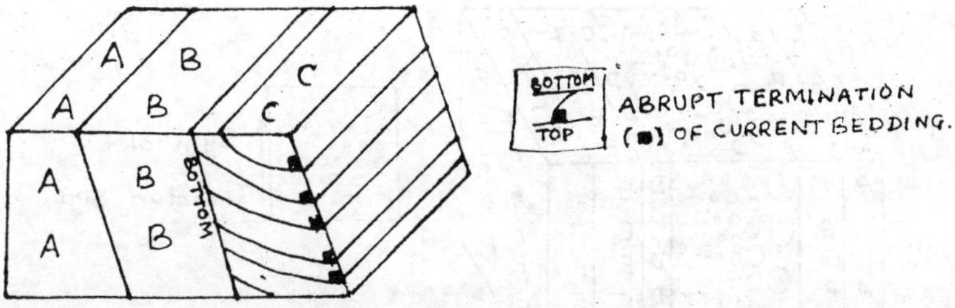

Fig. 19a. Disposition of the current bedding when the beds (rocks) are dipping. Beds A, B, C dip towards the right hand side. The current bedded structure is developed in bed C. The abrupt termination (which is also called abrupt truncation) is on the right hand side, and is not between the beds B and C. The top of the bed C is therefore surmised to be towards the right hand side i.e., in the direction of the dip of the beds. Therefore the bed A is considered to be oldest, while C to be the youngest one. Truncation is indicated by black squares in the figure.

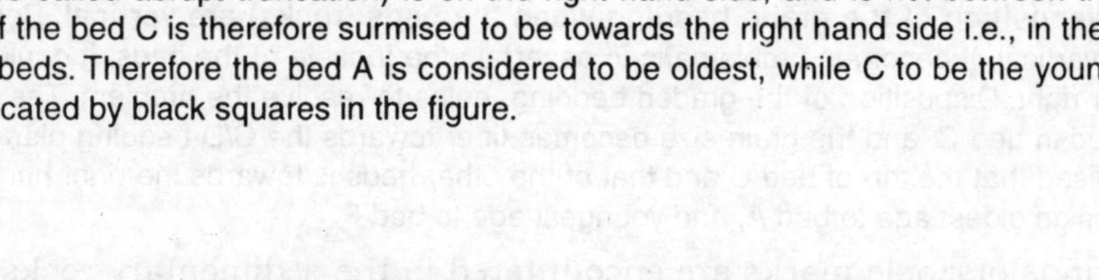

Fig. 19b. Disposition of the current bedding when the beds (rocks) are vertical. When the beds are vertical, it becomes problematic to ascertain the topside of the beds, because it could be on any oneside. The disposition of the current bedded structure, helps to solve the problem. This structure developed in bed B indicates that the truncation is between beds A and B, and not between B and C. Therefore the top of bed B is to be surmised to be towards the left hand side. Hence the bed A considered to the youngest and C to be the oldest one. In Fig. 19a, the ages assigned to beds A and C are just the opposite. Truncation is indicated by black squares in the figure.

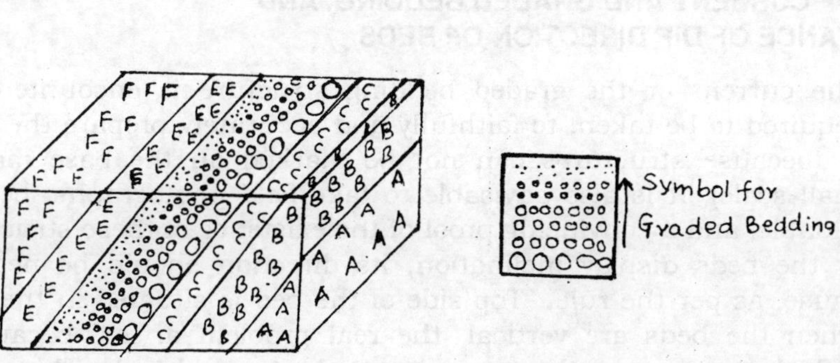

Fig.19c. Disposition of the grade bedding when the beds (rocks) are dipping. Beds A, B, C, D, E and F dip towards the left hand side. Graded bedding is developed in bed D. Note that the grain size becomes fine toward D/E bedding plane. Therefore it is to be surmised that the top of bed D is in the direction of dip of the beds. Hence it is further surmised that bed A is the oldest, while bed F is the youngest one.

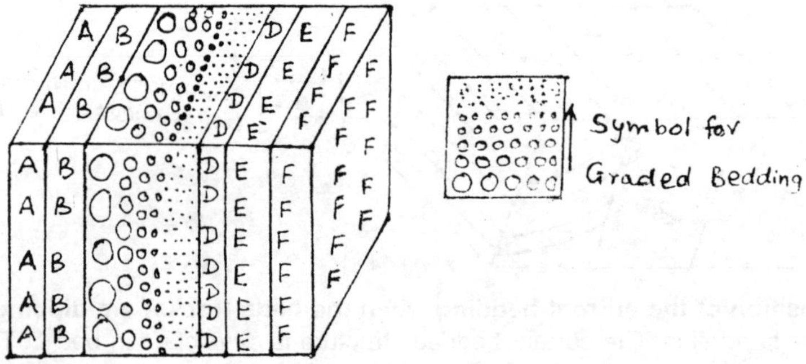

Fig.19d. Disposition of the grade bedding when the beds (rocks) are vertical. Beds A, B, C, D, E and F are vertical. It becomes problematic to ascertain the topside of the beds, it could be on any one side, left or right. Disposition of the graded bedding, helps to resolve the problem. The graded bedding is developed in bed C, and the grain size becomes finer towards the C/D bedding plane. Therefore it is to be surmised that the top of bed C and that of the other beds is towards the right hand side. It further helps to assign oldest age to bed A, and youngest age to bed F.

Two kinds of ripple marks are encountered in the sedimentary rocks. The current ripples are asymmetric in attitude, because these are produced by the flowing river water. The second variety is called as the "oscillation ripples". These are produced by the "up and down movement" of water, and these ripples therefore become symmetric in form. The difference between the two varieties is presented in Figs. 17 a, b.

From the trend of the axes of the current ripples, the direction in which the sediments were brought into the basin, can be established. If the axes be trending in a NE-SW direction, the sediments might have travelled in an easterly, or in a westerly direction. Therefore, the side on which the ripples have a steeper inclination, is required to be observed, and that gives the actual direction of transport of the sediments (Fig. 17b). The oscillation ripples are symmetrical in shape, and produce a concave surface pointing upwards (Fig.17a). This feature is made use of in establishing the top side of the bed (rock). This aspect has been shown in Figs 18a,b,c. The details of establishing the top of the bed, has been described by Gokhale (1993).

THE ROLE OF CURRENT AND GRADED BEDDING, AND THE IMPORTANCE OF DIP DIRECTION OF BEDS

Whenever the current or the graded bedding structure is encountered, extreme caution is required to be taken, to faithfully and accurately prepare the sketches on bigger scale (because structures can not be marked on the base map which is having a small scale). It is also advisable to take field photographs, because these provide a valuable, and a convincing proof of the existence of these structures.

Wherever the beds display inclination, its direction should be necessarily recorded, because, as per the rule, "Top side of the bed is situated in the direction of dip". But when the beds are vertical, the real problem arises, because, the dip direction could be on "one of the two sides of the vertical bed". This ambiguity is resolved with a reasonable degree of accuracy, from the disposition of the current bedding, the graded bedding, and the oscillation ripples, if registered in the rocks. The procedure has been elaborately described by Gokhale(1993). However for a ready appreciation, it has been briefly described below.

Locate the side on which the off-set beds" truncate abruptly" (in the case of the current bedding), or the direction in which the grain size becomes fine (in the case of the graded bedding), or the side on which the ripple is concave (in the case of the oscillation ripples). Then the side on which the off-set beds truncate abruptly, or the grain size becomes fine, or the ripples are concave, that side marks the top of the bed. These features are shown in Figs. 18 a,b,c and 19a,b,c,d.

Therefore, while mapping the sedimentary rocks, it is also necessary to devote attention to the current bedding, the graded bedding, and to draw separate sketches, besides taking field photographs of these structures.

The boundary of the sedimentary rocks with the igneous, or the metamorphic rocks, is required to be traced and mapped. Such a procedure alone will enable to describe the limit and the shape of the basin into which the sediments were laid. It is generally observed that the sedimentary basins are more long than wide, and possess an elongated form. Thus the famous "Tethys" which gave rise to the Alpine-Himalayan mountain ranges, is several hundred kilometers long, but the width is considerably small. The Kaladgi sedimentary basin is about 230 km. long, but the width is considerably less. The features to be emphasized is that the shapes of the sedimentary basins have been established, only after a careful, faithful, and accurate mapping of the sedimentary rocks.

The indicators of depth of a basin (and therefore the thickness of the sedimentary rocks comprising the basin) are primary structures namely, the ripple marks, the mudcrakes, and rain prints. These structures reflect a shallow depth of water, and in turn that of the basin. Therefore, while conducting mapping exercise, the presence of the said structures, and their disposition in the area, are required to be recorded.

The sedimentary rocks help to establish the climate under which the sediments were derived. The "Millette seed sands" in a sandstone, indicate an arid climate. "Facetted pebbles, unassorted pebbles, extremely large sized boulders, undecomposed laths of felspars, varved clays"' all of these e features, indicate a cold climate to be prevalent, at the time of disintegration and the decomposition of the parental rocks, and transportation and deposition of the products of weathering. Therefore while mapping, one should look out for the aforesaid features, in order to get an idea about the palaeoclimate of the region.

METHODOLOGY FOR METAMORPHIC ROCKS

These rocks share the characters of the sedimentary and the igneous rocks, because, these are derived from them. Therefore, to some extent, the features of the field techniques adopted for mapping the igneous and the sedimentary rocks, are also adopted to the metamorphic rocks too. However, there are certain petrographical, and the structural elements that are the monopoly of the metamorphic rocks. Those elements must be taken into consideration, while carrying out mapping, in a metamorphic terraine. The details have been elaborated in the following paragraphs.

The uniqueness of the metamorphic corks is their distinctive mineralogy and the textures. These rocks are recrystallised and reconstituted from the pre-existing sedimentary and/or the igneous rocks. The source of heat is required to be located, which had brought about the recrystallisation (metamorphism) of the pre-existing rocks. While dealing with the igneous rocks, it was said that the magma intrudes into the crustal rocks. As the magma starts crystallising, heat is given out, which in turn is transmitted into the rocks surrounding the magma. The outcome of such a process, is the recrystallisation of the country rocks. Obviously and logically too, rocks in the immediate vicinity of the magma, will receive maximum heat energy, while those located farther away from the magma, will receive feeble heat energy. The recrystallisation of the country rocks, will be therefore intense (highest degree) at the contact, while it will decrease away from the intruded igneous body (magma). Thus a gradation or zones of metamorphic rocks will be formed. This is spóken off as the "zones or the aureoles of contact metamorphism".

The objective of the geologist is to reconstruct the past "geological events" and this is applicable to the metamorphic rocks too. The grade or the degree (intensity) of metamorphism is established from the kind of the minerals and the texture developed in the rocks. It is also apparent that the geologists should be able to identify the minerals and the textures, during the field mapping exercise. The metamorphic minerals are either very hard like, the garnets, the corundum, the staurolite, or are very soft like, steatite, talc. Majority of the metamorphic minerals possess a platy or a prismatic habit, like the minerals staurolite, the kyanite, the glaucophane, the tremolite. Kyanite, besides being platy in habit, it possesses an unmistakable blue colour, and a hardness varying along and across the length (plate) of the mineral.

The metamorphic rocks alone possess the schistose and the gneissic structures, and the slaty cleavage. These textures and structures are readily and easily observable in the field. Therefore the metamorphic rocks could be easily identified during the field mapping exercise. However difficulty is faced while identifying the minerals in the field, and therefore the degree of metamorphism undergone by the rocks, can not be established during the field mapping exercise. However, with experience, the minerals indicating the degree of metamorphism, could be identified in the field, with a fair degree of accuracy. This should be later confirmed and/or rectified through the laboratory investigations (microscopic examination of the thin sections of the rocks) of the rock samples.

There is a good deal of correspondence between the actual parental sedimentary rock, and the metamorphic rock derived from it. This has been shown below.

SEDIMENTARY ROCK	**METAMORPHIC ROCK**
Sandstone	Quartzite
Shale	Slate, Phyllite, different Schists
Limestone	Marble

With the participation of the felspars and biotite, a gneissic texture is developed. The gneissic rocks lack the "cleavage or schistocity', and its place is taken up by the development of the "foliation". Thus the presence of the slaty cleavage, the schistocity or the foliation, is a reliable indicator of the presence of the metamorphic rocks in the area taken up for the mapping programme. Additional clues like the diagnostic metamorphic minerals, should also be thought of. The sedimentary rocks being bedded formations, the bedded nature is retained even after the metamorphism. This feature is very useful in deriving the nature (composition) of the parental rocks that had undergone metamorphism. The igneous rocks are not bedded ones, and these therefore can not give rise to bedded structures, after their metamorphism. The point to be emphasized is that during mapping, features that are helpful in deciphering the character of the parental rocks that had undergone metamorphism, should be observed and noted.

As already noted, metamorphic recrystallisation requires a source of heat. Two main sources have been proclaimed. A large sized igneous body will be capable of developing aureoles or shells of recrystallised country rocks around it, the intensity of metamorphism fading or decreasing away form the magmatic body. This process gives rise to the contact metamorphism, but the area affected by it, is relatively small. The second source of heat is the general rise of temperature with depth, in the crust of the earth. This situation also can bring about metamorphism of the rocks at depth. The area brought under the influence of this source of heat, is quite extensive, and the resultant metamorphism is therefore called as the "Regional Metamorphism". Therefore, while conducting the mapping of the metamorphic terraines, the extent to which the metamorphism has spread, should be accurately and carefully observed, because this alone helps to establish the type of metamorphism, whether the contact, or the regional one.

There is yet one nore kind of metamorphism, and it is associated with the major faults. It is called as the "Catalclastic Metamorphism". The products of such a metamorphism are the breccia, the cataclasite, the pseudotachylyte, the gouge, the mylonite, the flaser rocks, the autoclastic conglomerate, and so on. Interesting point is, that many times, the faults are detected through the existence of one or more of the said products. Displacement or throw can be observed in the case of the sedimentary rocks. But in the case of the igneous rocks, this criterion can not be applied, as these rocks are not bedded formations. The breccia and the conglomerate, are sedimentary rocks, but these could be produced during the process of faulting. Therefore the said structures should be carefully observed, as these might lead to the establishment of the cataclastic metamorphism and/or the existence of a major fault. It is therefore important to remember that there are "breccia and conglomerate" of sedimentary as well as of cataclastic (faulting) origin.

SUMMARY

From the foregoing account it is quite apparent that the methods of mapping the igneous, the sedimentary, and the metamorphic rocks, are different. But in all of them it is stressed that, extreme care is to be taken to inspect, each and every exposure of rock, in details. If necessary, separate sketches should be drawn on bigger scales, to document the important features, along side with taking field photographs. The vertical sections, wherever available in the field (the mines, the quarries, the road/railway cuttings, the gorges) should be carefully studied, and these should be drawn on bigger scales. Such sections go a long way in understanding the continuation or otherwise of the petrographical, as will as the structural features, at depth. The sections drawn in the laboratory from the geological maps, are expectations of the behaviour of the structural and the petrographical features, when continued (traced) at depth.

CHAPTER 5

Petrographical Field Features of Rocks

INTRODUCTION

In this chapter, it is intended to help the geologists in identifying the different rocks, during the process of mapping, with a fair degree of accuracy. This will help them not to miss collection of any sample whose composition is different from those already collected. However, the identification made during the field work should be confirmed and/or rectified through the microscopic examination of the thin sections prepared from the samples of rocks. The salient petrographical field features of the extensively and frequently encountered igneous, sedimentary, and the metamorphic rocks, have been described in the following paragraphs.

IGNEOUS ROCKS

These rocks occur as intrusives and as lava flows. The rocks belonging to the intrusive category are the syenite, the gabbro, the granite, the aplite, the pegmatite, the dolerite and basalt. The basalt also is the most preponderant rock amongst the category of the lava flows. Grain size holds the clue to distinguish a sample collected from an intrusive, from that collected from a flow. Granite, syenite and gabbro are the abyssal intrusive rocks i.e., these rocks are formed at great depths in the crust of the earth. These rocks are therefore very coarse grained. The granites, the syenites, the gabbro, the aplites, and the pegmatites, are seldom fine grained rocks. The dolerite and the basalt are formed at the intermediate depths in the crust of the earth, and these are catigorised as the hypabyssal igneous rocks. The aplite and the pegmatite occur as dykes or as veins. A feature of interest is, that the dykes of dolerite and basalt, are considerably fine grained, compared to the grain size displayed by the dykes and the veins of pegmatites and the aplites. The dolerite and the basalt are the basic rocks (like the gabbro), but the dolerite is coarser in grain size than the basalt. Therefore in the field, a dark looking dyke rock, is to be named as a dolerite (if coarse in grain size), or as a basalt (if it be of a finer grain size). The identification is to be confirmed through the examination of the thin section of the sample. A dolerite displays an ophitic texture, while a basalt typically displays an intergranular or intersertal textures, which is also described as the "basaltic texture".

The basalts predominantly occur as lava flows, but as already noted, these rocks also occur as dykes. The distinction is made by the presence of the "vesicular structure" which is invariably possessed by the flows of the basalts. This criterion has been successfully applied by Hegde, G.V. (1984) while mapping the eastern part of the Kaladgi basin, which exposes sedimentary rocks. He came across a number of boulders of basalt amidst extensively developed beds of the sandstones. However these boulders did not display any sharp contact with the beds of the sandstones. It therefore became problematic to categorise the boulders to formulate a dyke, because the boulders did not display any trend. On closer observation it was noticed that the rocks (boulders) possessed vesicularity. Hegde, G.V. (op.cit.) therefore categorised the boulders to formulate a flow, and not a dyke, Thus acute observation during the field mapping exercise helps to find solution to the problems cropping up during the geological investigations.

The *granite* is essentially a light coloured rock, either pink or light gray in colour. It is composed of felspars and quartz, as the essential minerals. Whenever porphyritic texture is developed, the potash felspars (the orthoclase or the microcline) compose the phenocrysts. The graphic and the rapakivi textures are the monopoly of this rock. These textures are readily and easily detectable in the field (from specimens collected during the field work), and confirm the rock possessing it, to be decidedly a granite. Biotite is the most commonly encountered ferromagnesian mineral in the granites. The granites invariably constitute bigger intrusions like the stocks, the bosses, and the batholiths.

Fig. 5a. Occurrence of orbicular structure in the gray granitic rocks exposed near Idgah Hill, Saundatti town, Belgaum district, Karnataka state. Five "orbicules" i.e, shells are clearly noticeable. Light and dark coloured shells alternate with each other. The light coloured shell is made up of felspars, while the ferromagnesian mineral forms the dark coloured shells. Each shell is about one cm thick. The ovoidal body is 12 × 7 cm in dimension. View of photograph looking S 80° W. Photo by the author

Fig. 5b. The orbicular structure i.e., the shells are not distinct as noticed in Fig. 5a, but numerous, elongated, ovoidal bodies (shells) are noticeable. The longer axes of the ovoidal bodies are nearly parallel to each other (as indicated by the handle of the hammer). Locality of the orbicular structures shown in Photos 5a and 5b, is about 2km. S 40°W of Saundatti town, Belgaum district, Karnataka state. View of photograph looking S 50°W. Photo by the author.

Fig. 5c. Three slabs seen in the photograph are of ideal and perfect orbicular structure developed in granites. The specimens are from the type-area of Finland. The name plates behind the slabs, are in the Finn language, and specimens have been displayed, outside of the museum of the Geological Survey of Finland, Helsinki. Note the perfect rounded, numerous light and dark shells, especially in the central slab. Photo taken in May-June 1965 by the author while in Helsinki.

The *syenite* is invariably pink in colour owing to the dominance of the potash felspars (the orthoclase or the microcline) in it. Quartz is generally absent, because, the syenite is less acidic than a granite. Quartz in fact is an essential mineral of the granites. Therefore by the megascopic (using a pocket lens) examination, if quartz be not detectable, chances are quite strong that the rock is likely to be a syenite. Instead of biotite, hornblende is the most preponderantly encountered ferromagnesian mineral in the syenites. Thus the dominance of the potash felspars, presence of hornblende, and the absence of quartz, are the salient distinguishing petrographical features of the syenites. These rocks do not constitute bigger intrusions like the stocks, the bosses or the batholiths, but these could be expected to occur in the close vicinity of the granites, than that of the gabbros.

It is difficult to identify a *Diorite* by the megascopic examination. However this rock typically displays the "orbicular structure" (Photos 5 a,b,c).

Therefore the presence of this structure, helps to recognise a diorite during the field studies. This rock is darker in colour than a granite, but is less dark than a gabbro, because, the diorite is of intermediate composition in respect of the content of SiO_2 in it. *Gabbro* is of a basic composition, and it is decidedly dark in colour, because of increase in the content of the ferromagnesian minerals, as compared to that occurring in the diorite. The gabbro possesses dark colour, also because of the absence of light coloured minerals like quartz and potash felspars in it. On weathering, the gabbroic rock developes a rough surface that resembles the skin of an elephant. It is in fact described as the "elephant skin weathering". Significant ferromagnesian mineral in the gabbros, is a pyroxene, which is identified by its rather more "equant grain size", unlike the platy or tabular habit of the amphiboles, or the flaky habit of the micas.

Difference between an *Aplite* and a *Pegmatite* is in respect of the grain size of the minerals composing these rocks. The pegmatite is extremely coarse grained rock, so much so that almost all of the minerals occurring in it, could be easily identified. On the other hand, an aplite is relatively, rather a medium grained rock, and it requires the assistance of a pocket lens, in the identification of the constituent minerals of the rock. There is yet another striking difference between the two rocks. A pegmatite contains mica (biotite, muscovite) as the chief ferromagnesian mineral. In fact, the pegmatites are ardently sought after rocks, because, deposits of mica are associated with it, besides a host of other minerals. The aplite is totally devoid of the ferromagnesian minerals, and it is therefore leucocratic (light coloured) in appearance. An aplite displays a sugary texture which is also called as the "aplitic texture".

A special mention is necessary for the dyke rock called as the "lamprophyre". This dyke differs from the other basic dykes (dolerite, basalt) in possessing a "pitch black colour", and the ferromagnesian minerals (augite, hornblende, biotite) forming the phenocrysts. The lamprophyre displays a "panidiomorphic texture".

NON-PETROGRAPHIC FIELD FEATURES OF THE IGNEOUS ROCKS

Veins of quartz are associated with the granitic rocks, while such veins are entirely lacking in the gabbroic rocks. The granites produce the "Tor or the watermelon topography". Both the basalts and the dolerites develop spheroidal weathering. Identification of the other igneous rocks like the diorite, the peridotite, the rhyolite, the andesite and so on, requires microscopic examination of the thin sections of the said rocks.

SUMMARY

The igneous rocks have a distinctive mode of occurrence, and petrographical features. Tor scenery, spheroidal weathering, flow and xenolithic structures, are some of the field features. These rocks display intrusive and the extrusive forms of occurrence. The vesicular, the rapakivi, the orbicular, the porphyritic, the chilled borders, are displayed only by the igneous rocks. These rocks cover extensive areas (except the dykes and the sills), and continue to extend to a considerable depth in the crust of the earth. Quartz and a variety of felpars (alkali felspars, plagioclase felspars, felspathoids) are considered in recognising varieties amongst the igneous rocks. The said two minerals are regarded as the essential constituents of the igneous rocks. Emphasis is therefore required to be laid on identifying these essential minerals during the field mapping exercise.

Considering the FOUR salient field features namely, i) mode of occurrence, and association, ii) grain size and texture/structure, iii) colour and iv) mineral composi-

Table 1. A schematic tabular field identification of igneous rocks

Mode of occurrence and asso-ciation	Large sized intrusions, extensive coverage in the area. Sometimes presence of xenotines, presence of pegmatite, aplite, quartz veins.	Medium to small sized instrusions, occurrence in the vicinity of gra-nites	Medium sized intrusions, or extrusions. Absence of peg-matite, aplite, quartz veins. Presence of basic rocks like flows of basalts, ultrabasic rocks like pyroxenite, du-nite etc.
Grain size, and texture, structure	Coarse to medium, Porohyritic, rapakivi, graphic, aplitic, por-phyry, pegmatic textures.	Medium Orbicular structure	Generally fine grained, but some show medium to coarse grain size. Flow,.scoriaceous, vesicular amygdaloidal, ropy structu-res/textures. Spheroidal weathering very common
Colour	Light gray coloured (leucocratic)	Intermediate coloured (mesocratic)	Dark coloured (melanocratic)
Mineral composition	Abundance of felspars and quartz + biotite	Abundance of potash felspars and plagio-clase felspars plus hornolende	Abundance of plagioclasses and pyroxenes
Rocks to be ex-pected	Acidic rocks (High SiO_2 content) Granite, Granodiorite, Pegmatite, Aplite, Rhyolite (if flow stru-cture, and fine grain-ed rock. This rock often shows bands). Pumice (scoriaceous structure), Obsidian (conchoidal fracture, smoky colour).	Intermediate rocks (Moderate content of SiO_2) Synite, Diorite, Dac-ite(if grain size be fine. Vesicularity may be seen. It is a volcanic rock)	Basic and Ultra basic rocks (Low content of SiO_2) Gabbro, Anorthosite, Dolerite, Basalt, Lampro-phyre, Pyroxenite, Peri-dotite, Dunite, Trachyte, Andesite (Last two are volcanic rocks)

tion, a schematic tabular classification has been prepared, to assist the geologists in the identification of the igneous rocks in the field. It is presented in Table 1.

SEDIMENTARY ROCKS

The rocks belonging to this group are few in number. The family names are given to these rocks, on the basis of the grain size, and not on the minerals composing them. The name *Rudite* includes conglomerates, grit and breccia. The *Arenite* covers a variety of sandstones. The *Argillites* include shale, argillite and mudstone. The *Calcareous* group (which is not distinguished on the basis of the grain size, as is done in the other three groups), has only two rocks in it, namely, a *Limestone*, and a *Dolomite*. Therefore the grain size possessed by the rocks is stressed more than the petrographical (mineral content) features, while distinguishing varieties within the sedimentary rocks. The petrography of these rocks is also very simple. Quartz, a variety of clay minerals, calcite and dolomite, are the predominantly encountered minerals in the sedimentary rocks. Quartz and calcite are easily identifiable (distinguishable) minerals through the contrast in respect of hardness (hardness of quartz is 7, while that of calcite is 3), and by the possession of a perfect rhombohedral cleavage by calcite, and absence of any cleavage in quartz. The clay minerals can not be identified merely by the microscopic examination of the thin sections prepared from the samples of argillites. The clay rocks require D.T.A., X-ray analysis, X.R.D. techniques. Thus it is clearly evident that the grain size holds the main clue to distinguish the sedimentary rocks during the field mapping exercise.

When the pebbles and the boulders are too large, a rock composing them is called as a *Boulder bed*. When the pebbles are smaller in size, and are rounded to sub-rounded in shape, the rock possessing them, is called a *Conglomerate*. In a *Grit*, the constituent grains are angular in shape, but the grains are smaller in size than those present in a conglomerate, but are coarser than those encountered in a sandstone. When the angularity of the grains is much pronounced, and the grains appear as "broken pieces of rock", such a rock is named as a *Breccia*.

A *Sandstone* is entirely make up of grains of quartz. When grains (laths) of felspars are present, alongside with those of quartz, such a rock is called as a *Felspathic Sandstones or an Arkose*. The argillaceous group displays fissility (a sort of cleavage). The *Shales and the Argillites* are very fine grained rocks. *Laterite and Bauxite* are often times associated with the shales and the argillites. The *Laterite* is reddish brown in colour, and it possesses a porous structure (which is in fact called as the lateritic structure). The *Bauxite* is typically light brown to white in colur. The *Limestones* often develop a brownish colour as a result of weathering. The brownish colur is due to the separation of ferric carbonate ($FeCO_3$) fraction held in a state of "solid solution" within $CaCO_3$. A *Dolomite* does not hold $FeCO_3$ in $MgCO_3$, and therefore, weathering of this rock does not produce a brownish colur on its surface. The brownish colouration on the surface of the limestones, is due to the oxidation of $FeCO_3$ to iron hydroxide, the latter giving rise to the rusty colour.

There is a confusion and a difference of opinion regarding the rock "*Graywacke*". Pettijohn (1964) has opined that "what is described as a graywacke, is in fact a breccia. It has a considerable resemblance to a sandstone except the gray colour, which is lacking in a sandstone". It is desired not to enter into any controversy, because, the "core intention" of the present book, is to guide the geologists in the identification and a systematic mapping of the rocks, that are dominantly encountered in the crust of the earth. It is therefore enough to note that a graywacke shares the features of a sandstone, as well as that of a breccia.

SUMMARY

The chief petrographic features of the sedimentary rocks are the grain size, presence of a cementing material that binds the constituent grains, and the presence of either quartz, clay minerals, calcite or dolomite. The sedimentary rocks are nearly of a monomineralic composition, and one or two of the minerals as noted above, compose these rocks.

METAMORPHIC ROCKS

These rocks are distinguished mainly on the basis of the structures possessed by them, as is done with the sedimentary rocks too. The metamorphic rocks being derived from the igneous, or the sedimentary rocks, some of the parental structural/textural features are often retained in them. Such structures are called as the "Palimpsest or the Ghost structures". Thus the bedded nature of the sedimentary rocks, is many times not destroyed during the process of recrystallisation. A good example is the rock "banded hematite quartzite". This was previously a layered or a bedded, siliceous-ferruginous rock. From the siliceous and the ferruginous layers, quartzite and hematite layers were respectively formed, as a result of the metamorphic recrystallisation. Besides possessing some salient structures and textures (like the slaty cleavage, the schistosity, the foliation), there are minerals which are the monopoly of the metamorphic rocks. The typical minerals are the garnets, the staurolites, the kyanites, the corundum, and so on. These two features are the main guide lines in distinguishing and identifying the metamorphic rocks in the field.

Two structures are dominantly encountered in the metamorphic rocks, namely, the schistosity and the gneissic foliation. The words "schist, and gneiss" are used as suffixes in naming majority of the metamorphic rocks. Thus there are chlorite schists, hornblende schists, mica-garnet schists, biotite gneisses, sillimanite gneisses, and so on. The schists and the gneisses are the most extensively developed metamorphic rocks. A schist is identified from the characteristic greenish colour and the presence of cleavage (called as the schistocity) in the rock. The schistocity is mostly due to the presence of platy/flaky minerals like the chlorite, the biotite and the hornblende. The schistose rocks are named on the basis of dominant ferromagnesian mineral composing them. It is therefore necessary to identify the ferromagnesian minerals, in order to name the schist. By and large, the schistose rocks are dark in colour, because of the dominance of the ferromagnesian minerals, over the quartz and the felspars. Therefore it may be said that the characteristic petrographical feature of the schists, is the cleavage (schistocity) and the greenish colur. The schists cleave (break or split) easily when hammered, unlike the gneisses.

The gneissic rocks display foliation. This feature differs from the schistocity in that, the platy, flaky ferromagnesian minerals (mica, amphiboles) are not confined to parallel planes. Further, the gneisses contain a considerable amount of the felspars and the quartz, in them. As a result of this, when a gneissic rock is broken with a hammer, it does not "cleave" cleanly into slabs or plates, as do the schists. The gneisses are light gray in colour. The nomenclature of these rocks is similar to that adopted with the schists i.e., the dominant ferromagnesian mineral is made use of in giving a name to the rock. Thus the names given are, biotite gneiss, hornblende gneiss, sillimanite gneiss and so on. Sometimes the entire composition of the rock (and not the dominant ferromagnesian mineral, as is the case with the schists), is taken into account, and the suffix of "gneiss" is added. This practice is followed whenever an igneous rock has been metamorphosed. Thus there are such names as a granite gneiss, a granodioritic gneiss, diorite gneiss, and so on. A gneissic rock wherein a resistant constituent mineral, is lenticular or oval in shape, is called as an "augen gneiss" (Photo 6).

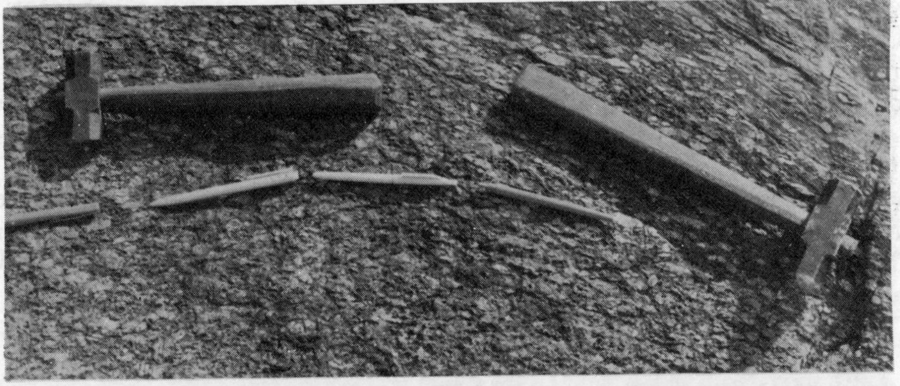

Photo 6: Exposure of an augen gneiss in the eastern vicinity of Akuthothapalli village, Mehboobnagar district, Andhra Pradesh. The handle of the hammer (the one on the left side of the photo) is parallel to the longer axes of the augens of pink coloured potash felspars. Note that there is dominance of felspars and quartz, over the ferromagnesian minerals, and therefore the rock is leucocratic. The rock displays foliation and not schistocity (Courtesy, Dr. D. Muralidharan).

The augen (meaning an eye-shaped body) is invariably made up of felspars. Gothe (1973) has described such rocks from Mundargi, Gadag district, Karnataka state.

There are some metamorphic rocks that require a special, and a separate mention. The *Amphibolite* is one such rock. It is in fact a hornblende schist, in respect of its mineralogical composition. The amphibolite also displays the typical schistose structure. Even then, some rocks are named as amphibolite, and not as hornblende schist. In the field however, it is difficult to make a distinction between an amphibolite, and an hornblende schist. The thin section examination of these rocks, is necessary to make a definite identification. *Granulite* is yet another metamorphic rock that needs a mention. As already noted, the metamorphic rocks are characterised by the schistose and the gneissic structures. However some display a granular structure, like that displayed by the granites, or by the sandstones. The constituent minerals of a granulite exhibit some degree of flattening and elongation, especially, quartz and the pyroxenes. Therefore, a granular texture with typical metamorphic minerals, is considered in naming these rocks as granulites. These reocks generally contain pyroxenes as the ferromagnesian mineral, and less frequently the amphiboles or the biotite. One such pyroxene granulite has raised a controversy in petrology, regarding its origin. The rock called as the "charnockite" is in fact a hypersthene granulite. The granulites are very useful in ascertaining the degree of metamorphism undergone by the rocks. These rocks are regarded to belong to the highest grade of metamorphism, and it is designated as the "granulite facies of metamorphism".

There are rocks called as the metadolerite, metabasalt, metagabbro, epidiorite, and so on. These are all metamorphosed basic igneous rocks, but these can not be identified by the megascopic examination, during the field mapping exercise. The thin section study of the said rocks reveals, that the original pyroxenes have been uralitised (i.e., converted into amphibole), and the plagioclase felspars have been saussuritised (i.e., converted to albite, with release of quartz etc.). A prolonged field experience alone will enable the geologists to identify the metabasic rocks (metadolerite etc.) in the field. Thin section study of such rocks is however necessary. The parental sedimentary and the igneous rocks, get metamorphosed into schists, gneisses and granulites. But the mode of occurrence of the parental rocks however, is many times not destroyed during the process of metamorphism. Therefore the field set up provides a vital evidence in respect of the nature of the parental rocks. One instance is cited to illustrate the importance of observing the field set up of the rocks, during the field mapping programme. Hegde, V.N. (1984) has reported an amphibolite dyke traversing the granites of Hampi area, Hospet taluka, Bellary district, Karnataka state. This rock has developed schistocity, yet its mode of occurrence shows an unmistakable cross cutting relation with the granites (Photo 7).

Photo 7: Amphibolite dyke exposed at about 2 km. WNW of Ramsagaram village, Bellary district, Karnataka state. It traverses gray granite gneiss. Note the distinct cross cutting relation with the granite gneiss. The dyke rock has developed schistocity, yet it shows typical discordant relation which is the characteristic feature of the intrusions. (Courtesy, Dr. V.N. Hegde, who is taking the trend of the dyke, with a clinometer compass).

Such a field set up helped Hegde, V.N. (op. cit.) to interpret that a basic dyke which had earlier traversed the granites, was subsequently metamorphosed into an amphibolite. It is to be remembered that a basic dyke can intrude, but not an amphibolite. Therefore the amphibolite under discussion, must have been derived from a basic dyke, through the process of metamorphism.

The "flaser rocks" are important both from the standpoint of metamorphism, as well as structure. A rock in which considerable differential movement has taken place, and lenticles (pieces) of relatively unaltered parental rock, are enveloped in a finely crushed material, is called as a "flaser rock". Such rocks are named after the mineralogical composition (petrography) of the unaltered lenticles. These could be

Photo 8: Exposure of unmistakable pillow structure developed in the schists of Attikatti-Mahalingpur area, Gadag district, Karnataka state. Note the development of skin for each pillow, and a soft contact between the adjacent pillows, and the concentration of vesicles/cracks along the border portion of the pillows. Also note the ovoidal shape of the pillows. These could be mistaken for the boulders of a conglomerate bed, and hence a very critical and a minute examination of the outcrops of rocks is essential (Courtesy, Dr. B.P. Waghamare).

granite, granodiorite, gabbro, and so on. Accordingly the rocks are named as "flaser granite, flaser gabbro" and so on. The other aspect of importance is that the occurrence of these rocks, mark the sites of extensive movement, and hence the existence of thrust or related faults.

A mention has been made to the "Palimpsest or the Ghost" structures, which are encountered in some metamorphic rocks. There are numerous ancient schistose rocks forming belts in India, like the Sandur schist belt, the Dharwar-Shimoga schist belt, the Gadag-Shrirangpattan schist belt, and so on. In some of these schist belts, schists possessing pillow- structure, have been described. Pillow structure is developed only in lava flows. Such flows later get metamorphosed into schists, but the pillow structure remains intact. This structure therefore belongs to the category of the "palimpsest": structure. Gokhale etal. (1981) have described the occurrence of the pillow lavas from the Gadag schist belt (Photo 8). Desai (1989) has reported the occurrence of pillow structure from the vicinity of Vadrahalli village, Bellary district, Karnataka state (Photo 9).

SUMMARY

The petrographical field features of the metamorphic rocks, are the typical minerals like the garnet, the kyanite, the staurolite, and structural features like the schistocity,

Photo 9: Pillow structure very well developed in the actinolite schists of Karadikolla region, Bellary district, Karnataka state. Length of the pillow (indicated by the scale) is 20 cm. Note the lenticular shape of the pillow. This is a "Ghost structure" in the context of the parental rock of the present day actinolite schist, is concerned. The outcrop documented in the photograph, exposes numerous pillows (Courtesy, Dr. H.D. Desai).

the gneissic foliation, the palimpsest structures, and so on. These rocks share the features of the igneous and the sedimentary rocks.

GENERAL SUMMARY

The petrographical field features of the igneous, the sedimentary and the metamorphic rocks, that are identifiable in the field (megascopically) and are commonly encountered, have been described above. There are a host of other rocks which do occur in the crust of the earth, but those are not easily identifiable in the field, because those require a microscopic examination of the thin sections prepared from the rock samples. To avoid complication and confusion in the mind of the geologists, such rocks have not been described in this book. However it is suggested that the samples should be collected even if the identification be not possible. At best it might result in duplication of the samples (because it might turn out to be similar in composition, to the one already collected). However, there is also the possibility of the sample to be totally different from those collected and identified in the field. This might lead to the discovery of a rock, so far not reported from the area by the earlier workers. More the number of samples, better it is to depict and understand the petrography of the area taken up for mapping programme.

CHAPTER 6

Structural Field Features of Rocks

INTRODUCTION

The structural features have been described in a separate chapter for the simple reason of not mixing up these with the petrographical features. It is a fact that in any outcrop of a rock, these two features occur together, and not independent of one another. The geologist therefore has to develop the vision of recognising both the features during the process of mapping. Some of the structural elements have been already noted like the bedding planes, the current bedding, the ripple marks (all from the sedimentary rocks), flow banding, ropy structure, xenolithic structure (all from the igneous rocks), schistocity, slaty cleavage., gneissic foliation (all from the metamorphic rocks). These were described in brief, in order to distinguish the igneous, the sedimentary, and the metamorphic rocks, from one another, in the field, during the field mapping exercise. There are several other structures that are needed to be observed and recorded and/or sketches to be drawn. These will be described in the following paragraphs.

IGNEOUS ROCKS

The structures possessed by these rocks are distinctly different from those encountered in the sedimentary, and the metamorphic rocks. During the act of intrusion of the magma into the country rocks, these latter rocks develop structures in them, in order to accommodate (create space) the intruding magma. The "domical or the arch structure" of the igneous body (rocks), is entirely registered in the layers or the beds of the country rocks (if these be sedimentary ones). Otherwise, the magma being in the liquid state, it alone cannot produce a domical form. Therefore the "laccolithic structure" is in fact produced in the sedimentary rocks (forming the country rocks), but the "laccolithic shape" is assigned to the igneous body (rocks), and it is deemed to be its mode of occurrence. It is important to note that the igneous rocks do not possess any bedding planes, and therefore, these rocks can not develop any inclined planes, like that associated with the sedimentary rocks. The structure of the "Qua Qua versal dip" displayed by some igneous intrusions, is in actuality the inclination developed by the invaded, bedded, sedimentary rocks. Unless the invaded country-rocks are made up of beds, the igneous intrusion by itself, can not produce the "laccolithic shape, or the Qua-Qua versal dip" structures.

In the case of the granites, the upward movement of the magma (which later gives rise to the granites), is determined from the disposition and the attitude of the "rupture structures" like the shears, the joints developed in the large sized intrusions (like the batholiths), and in the intruded country rocks. This feature has been designated as "Granite Tectonics" by Cloos (1939). Gokhale (1996) has given a

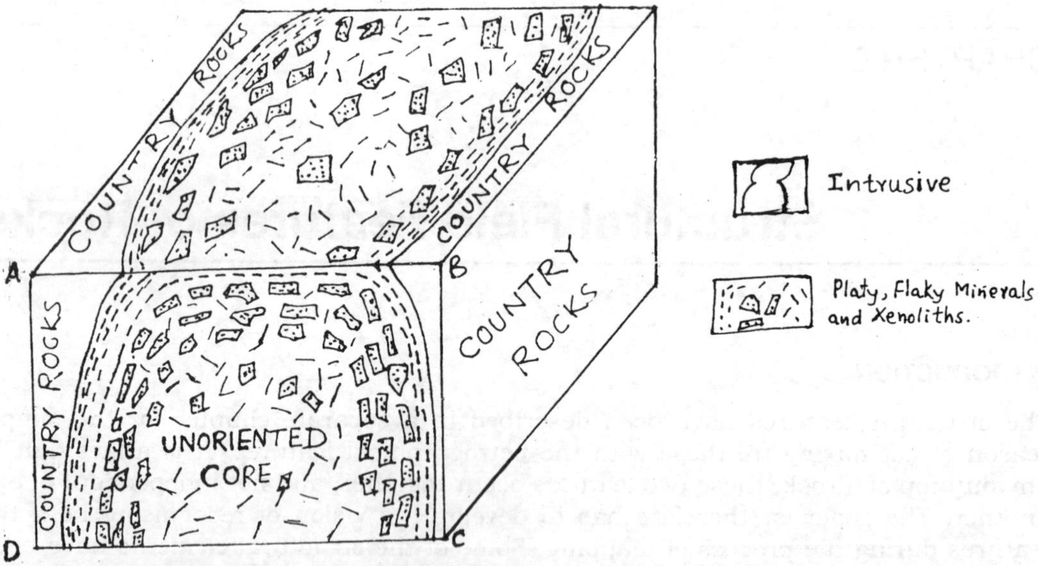

Fig. 20a. Structural features of large sized plutonic intrusives. The foliation and orientation produced by the platy/flaky minerals and the xenoliths, imparts a "domical or an arch structure" when a vertical section of large sized intrusive is examined. ABCD is one such vertical section. After Fig. 222, Theory of structural Geology, Gokhale 1996).

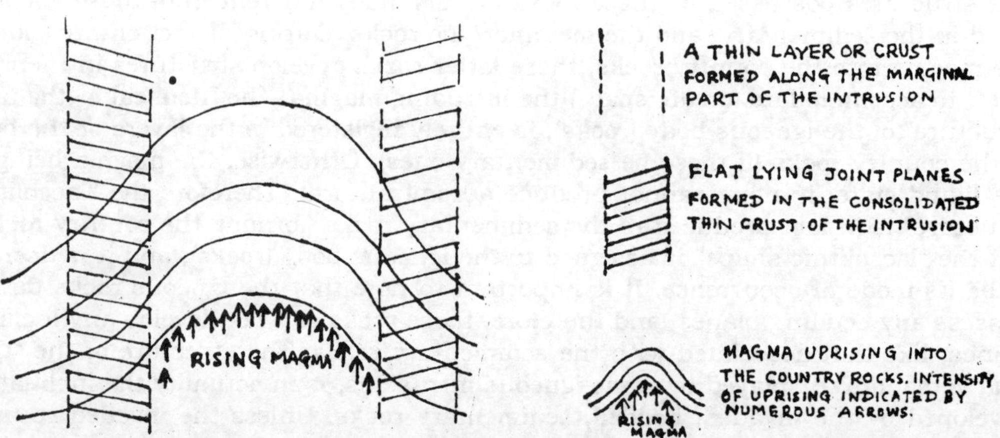

Fig. 20b. Initial stage of magma rising vertically upwards. The magma rises vertically upwards, quite fast, and pushes its way against the walls of the magma chamber (composed of country rocks). During this stage, a thin crust is formed along the marginal part of the intrusion. Joints are formed in the consolidated marginal part of the magma, and these joints possess a low inclination. Such joints are called as "Cross Joints".

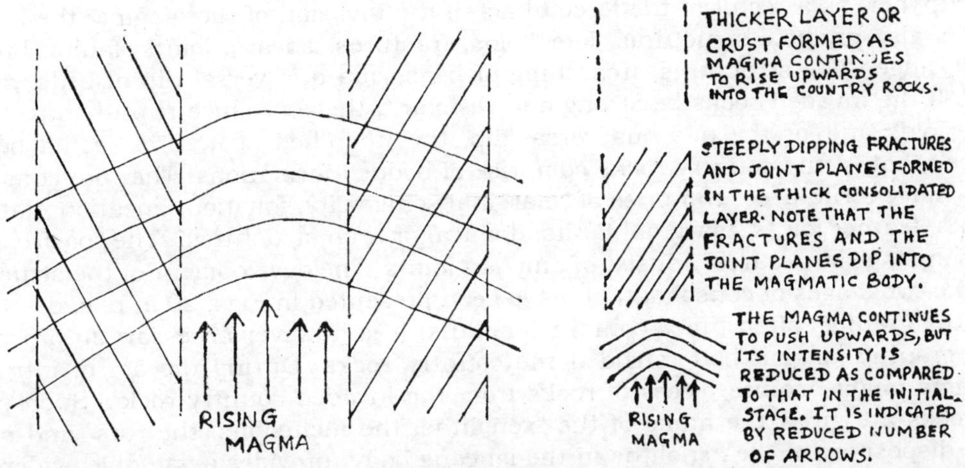

Fig. 20c. Middel stage of magma rising vertically upwards. The magma rises upwards but with a reduced intensity, because a thicker crust is formed over the rising magma. As a result of the magma pushing upwards, steeply dipping fractures and joint planes are formed in the consolidated part of the magma. Some of the joint planes are utilised as fault planes. These fault planes give rise to "upthrusts". The high angle joint planes are called as "Feather Joints".

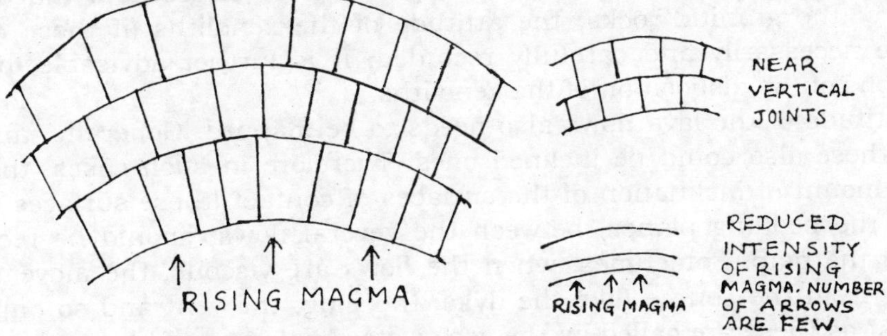

Fig. 20d. Final stage of magma rising vertically upwards. This is the final stage in the upward movement of the magma. Practically almost the entire magmatic body has crystallised, and the remainder magma rises up, very feebly. At this stage, almost vertical joints are developed, and these fan out in the magmatic body. These joints are described as the "Fan Joints".

Fig. 20a,b,c,d. In Fig. a, a vertical section of a large igneous intrusion is presented. The curved lines in Figs. b,c,d indicate the intensity of upward movement of the magma. Note that the curvature is steep in (Fig.b) because the intensity is maximum, the curvature is gentle (Fig. c) because the intensity is moderate, while the curvature is extremely gentle (Fig. d) because the magma pushes its way upwards very feebly.

detailed account of the various aspects of the "Granite Tectonics". A brief account of it however, is given here to enlighten the salient aspects. Structures are developed in the intrusion, as well as, in the invaded country rocks. In the intrusion, the structures are developed in two stages, namely,

(i) pre-consolidation stage, and
(ii) the post-consolidation stage.

Further, as a result of the forceful act of the invasion of the magma, the country rocks also produce structures, like "folds, fractures, shears, joints, faults, breccia-tion, augen-like structures, uparching of beds, qua qua versal dip of beds, and so on". If the invaded rocks be strong and resistant, these produce rupture structures. The folds and/or the qua-qua versal dip, the uparched structures, could be produced, only if the country rocks comprise of bedded formations. The "pre-consolidation stage structures" comprise of "marginal schistocity, foliation, lineation, flat lying joints, feather joints, marginal faults, domical or arch structures". The "post-consolidation stage structures" consist of only fan joints. The development of the structures in the two stages of consolidation, have been presented in Figs. 20 a, b, c, d.

The granites often times contain xenoliths. It is however necessary to prove that the granitic magma had intruded the country rocks. During the act of intrusion, magma might acquire pieces of rocks from the invaded country rocks. Such pieces of rocks are given the name of the "xenoliths, the inclusions, the rats" and so on. The disposition of the xenoliths in the igneous body, provides a valuable evidence to establish the mode of origin of the granitic rocks. If the disposition of the xenoliths be irregular and haphazard, it is surmised that the magma did exist, and it forcibly invaded the country rocks, in order to create "room" for itself in the crust of the earth. If on the other hand, the disposition of the xenoliths, displays an orderly and a systematic pattern, and if the longer axes of the xenoliths, display parallelism between themselves, such a disposition indicates the possibility of a "non-magmatic origin" of the granitic rocks. Hence the point to be stressed is that during the mapping of the granitic rocks, the attitude of the xenoliths (if these be present) should be necessarily and carefully recorded. It is further advisable to take field photographs of the disposition of the xenoliths.

The attitude of the lava flows also needs to be mapped. Generally, the flows are flat, but these also could be inclined ones. Therefore in such cases, the direction and the amount of inclination of the surfaces of contact (these surfaces can not be called as the bedding planes) between the several flows, should be recorded and shown on the map. Sometimes, when the flows are viscous, the movement of the lava away from the source (like the dyke, the plug, the vent and so on), produces "flow lines" as these are called by the geologists. Analogy could be given of a flowing mass of tar. Tar as it flows out, its surface produces "curved boundaries". Likewise, the irruption of lava from a vent takes place in pulsations. Each pulsation causes curved surfaces or wrinkles (called as the flow lines) on its surface. Such a study of the flow lines, is especially useful, when the lava flows cover extensive area, like the Deccan Traps of India. The mode of formation of the Deccan Traps, whether from a crater, or from a fissure, it could be possibly inferred from a study of the flow lines. It is logical to expect the lava to "well out" quietly from a fissure, than from a crater. Based on the surficial appearance (which tentamounts to a structural and not to a petrographical studies), "aa, pahoe-hoe, ropy lavas" are to be distinguished. These are generally not mapped, but are recorded and documented in the form of field photographs.

SEDIMENTARY ROCKS

The dominant primary features possessed by these rocks are the bedding planes, the surface of contact with the older country rocks, this surface being called as the "plane of inconformity". Therefore the direction and the amount of dip of these

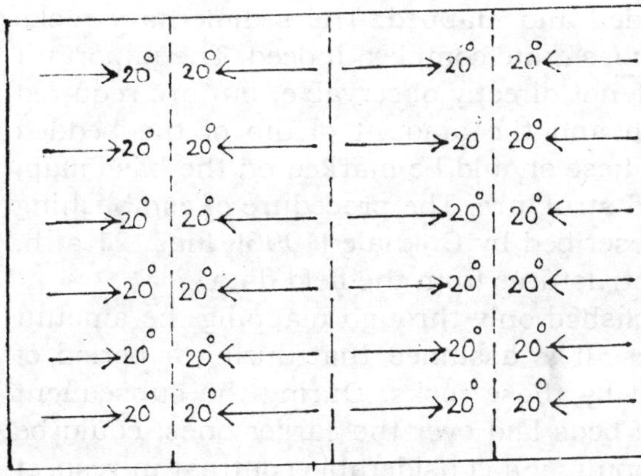

Fig. 21a. Symmetrical folds. Two synclines and one anticline established through recording and plotting direction and amount of dip of beds. Dashed lines mark the axes of folds.

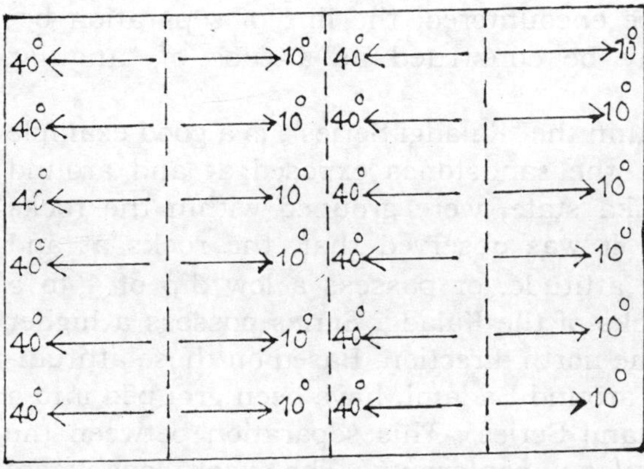

Fig. 21b. Asymmetrical folds. Two anticlines and one syncline established through recording and plotting direction and amount of dip of beds. Dashed lines mark the axes of folds.

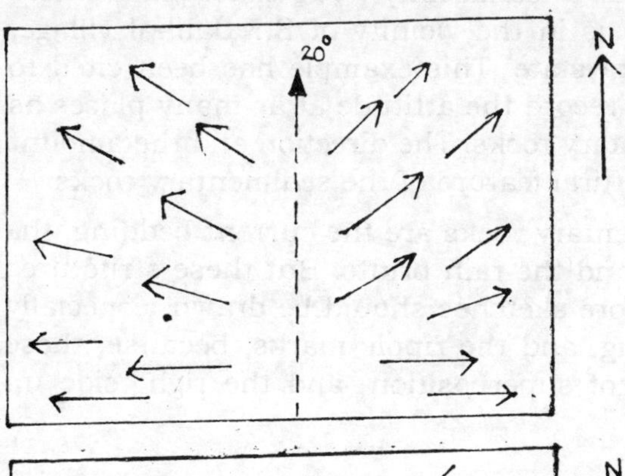

Fig. 21c. Plunging anticline. The plunging nature of the fold is established by recording and plotting the amount and direction of dip of beds. The dashed line marks the axis of the plunging anticline. The thick arrow indicates the direction of plunge. The amount of plunge is either equal to or less than that of the limbs of the fold. The limbs possess a dip of 40°, the amount of plunge is 20 degrees.

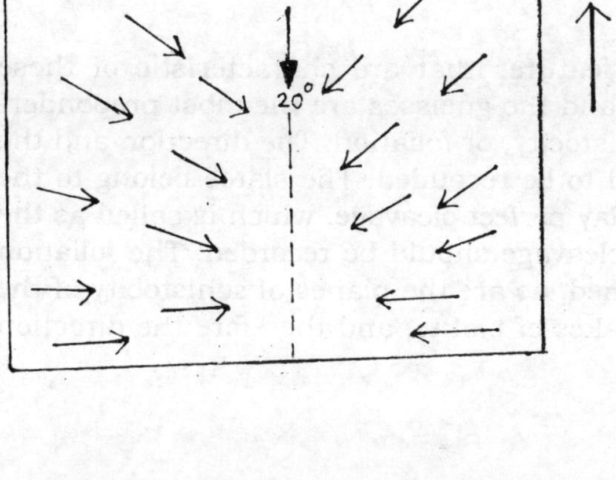

Fig. 21d. Plunging syncline. The plunging nature of the fold is established by recording and plotting the amount and direction of dip of beds. The dashed line marks the axis of a plunging syncline. The thick arrow indicates the direction of plunge. The amount of plunge is either equal to or less than that of the limbs of the fold. The limbs possess a dip of 40°, the amount of plunge is 20 degrees.

planes/surfaces, are required to be recorded and mapped. The sedimentary rocks are often thrown into folds, which are at times quite complex, indeed. The important aspect is that the majority of the folds, are not directly observable, but are required to be established, by noting the direction and the amount of dip of the bedded formations, at as many sites as possible. These should be marked on the base map, which might reveal the presence of a folded structure. The procedure of establishing the folded nature of the rocks has been described by Gokhale (1996). Figs. 21 a, b, c, d, illustrate the derivation of the folded structures from the field data.

The angular unconformity can be established only through mapping the amount and direction of dip of the bedding planes. It is assumed that during a period of sedimentation, certain attitude is attained by those rocks. During the subsequent event of sedimentation, the attitude of the beds laid over the earlier ones, could be different. Hence during the process of mapping, if a considerable contrast in respect of the amount and/or direction of dip be encountered, the line of separation between the two different attitudes, may be construed as a case of "angular unconformity".

The creation of the "Badami Series" within the "Kaladgi Series", is a good example of the angular unconformity. Previously, the sandstones exposed at and around Badami town, Bagalkot district, Karnataka state, were grouped within the rocks comprising the Kaladgi Series. Later on it was observed that, the rocks at and around Badami, are either horizontal in attitude, or possess a low dip of 4 to 5 degrees, generally due north. But the rocks of the Kaladgi Series possess a higher amount of dip, though the beds dip due the north direction. Based on these attitudinal differences, the rocks exposed at and around Badami, have been grouped into a separate serics, and named as the "Badami Series". This separation between the rocks is effected on the basis of the "angular unconformity". The exact place where the angular unconformity can be observed is in the vicinity of B.N.Jalihal village, Badami taluka, Bagalkot district, Karnataka state. This example has been cited to emphasize the point that, it is necessary to record the attitude at as many places as possible, during the mapping of the sedimentary rocks. The direction and the amount of dip, therefore constitutes the major structural feature of the sedimentary rocks.

The other primary features of the sedimentary rocks are the current bedding, the ripple marks, the sun or the mud cracks, and the rain prints. But these structures can not be shown on the base map. Therefore sketches should be drawn, especially for the current bedding, the graded bedding, and the ripple marks, because, these are very useful in establishing the "order of superposition, and the right side up position" of the sedimentary rocks (beds).

METAMORPHIC ROCKS

These rocks possess very few structural features that are characteristic of these rocks alone. As already noted, the schists and the gneisses are the most preponderant rocks, and these possess planes of schistocity, or foliation. The direction and the amount of dip of these planes are required to be recouded. The slates belong to the family of the schists, and these slates display perfect cleavage, which is called as the "slaty cleavage". The attitude of the slaty cleavage should be recorded. The foliation planes of the gneisses, are not so well defined, as are the planes of schistocity of the schists. The gneisses commonly contain flakes of biotite, and therefore the direction

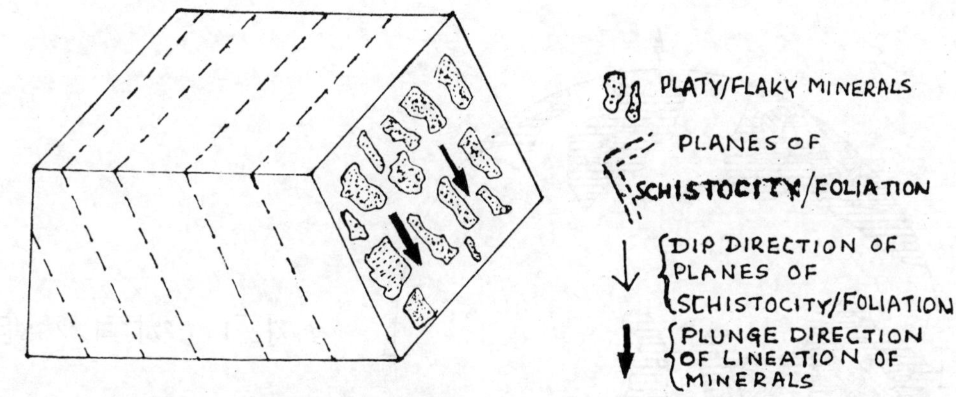

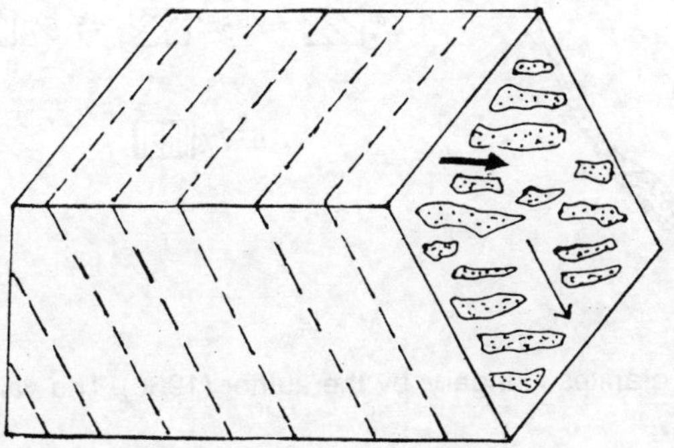

Fig. 22. (a) The direction of dip of Schistocity/ foliation (thin arrow), is coinciding with the direction of plunge (thick arrow) of lineation of minerals. (b), the direction of plunge of lineation of minerals (thick arrow) is at an angle to the direction of dip of schistocity/foliation (thin arrow). The dashed lines indicate the planes of Schistocity/ foliation.

and the amount of dip possessed by the flakes of biotite, are to be recorded. The presence of plates and the prisms of hornblende in the gneisses, might develop lineation, because this mineral is more long in one direction. The minerals actinolite and staurolite also produce lineation. The direction and the amount of dip of lineation, is called as the "bearing and the plunge of lineation". Sometimes the laths of felspars also produce lineation. Likewise the rock called as the "augen gneiss" often develops lineation. Therefore the "bearing and the plunge of lineation" should be recorded., The relation between the dip of schistocity/foliation, and the plunge of lineation, is presented in Figs. 22 a, b.

STRUCTURAL FEATURES COMMON TO ALL ROCKS

So far the structural features confined to the igneous, the sedimentary, and the metamorphic rocks, have been described separately. Besides these, numerous other structural features are developed in the rocks irrespective of the category to which they belong. Those structures are the joints, the shears, the faults, and so on. What is to be recorded of these structures, has been described in the following paragraphs.

JOINTS

n the case of joints, the amount and the direction of dip of the joint-planes, are equired to be recorded. Ideally, the quarries are best suited for recording the ttitude of the several joint-planes. The readings of joints from one quarry should not be mixed up with those collected from another quarry, eventhough the rock xposed be same. It is suggested that in order to comment on the pattern of joints,

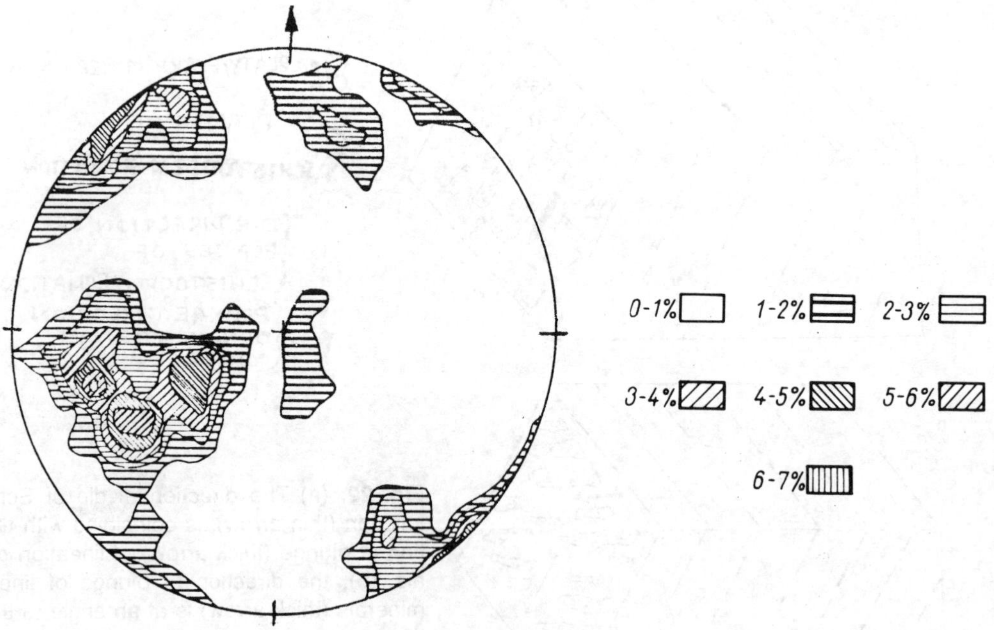

Fig. 23a. Stereogram of joints recorded in Velence granites Hungary, by the author (1966). The values indicate the density of joints developed in one quarry.

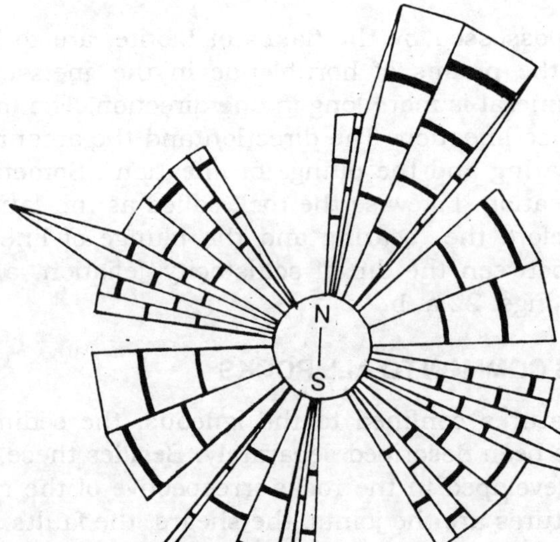

Fig. 23b. Rosette diagram of directions of shearing. Developed in granites of Andurla, Vengurla area, Ratnagiri district, Maharashtra state (After Deendar, 1982).

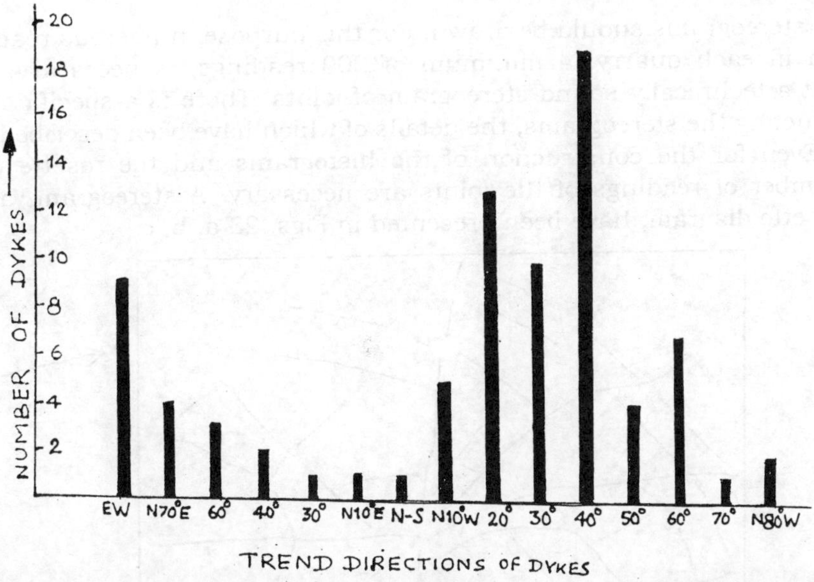

Fig. 23c. Trend directions of dykes. Histogram of basic dykes traversing granites adjacent to the northwestern part of the Gadag Schist belt, Gadag district, Karnataka state (Data from Waghmare, 1990)

Photo 10: Exposure of radial joints. The several directions are N40°E, N45°E, N85°E, N10°W, N58°W, N60°W, and N85°W. Note that at the intersection of the radial joints, the rock is broken. The photo also shows presence of cross joints, inbetween two adjacent radial joints (left hand central part of the photo). Two such cross joints are prominently seen-one trending in N40°E-S40°W direction, and another in N60°E-S60°W direction. The radial and the cross joints together produce a "cobweb like structure". Location of this exposure is at a distance of 2 km. NE of Gorvankolla village, Saundatti taluka, Belgaum district, Karnataka state (Courtesy, Dr. G.S. Pujar).

separate stereograms should be drawn. For this purpose, numerous readings should be taken in each quarry, a minimum of 200 readings, is necessary, in order to construct a technically sound stereogram of joints. There is a specific methodology of constructing the stereograms, the details of which have been described by Gokhale (1996). Even for the construction of the histograms and the rosette diagrams, a large number of readings of the joints are necessary. A stereogram, a histogram and a rosette diagram, have been presented in Figs. 23 a, b, c.

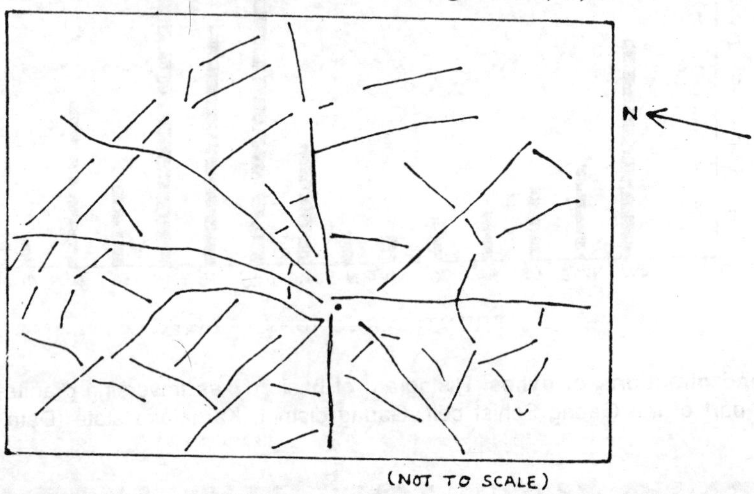

Fig. 24. Cobweb-like structure. The cobweb-like structure is produced by the radial and the cross joints. Trend directions of joints are N40°E, N45°E, N85°E, N10°W, N58°W, N60°W, and N85°W. This pattern is exposed 2 km. NE of Gorvanakolla village, Saundatti taluka, Belgaum district, Karnataka (After Dr. G.S.Pujar, 1989).

At times the joints display a radial pattern or a cob-web like pattern, and these can be recognised, if carefully observed during the field work. At such places, field photographs should be necessarily taken, and the readings of joints pertaining to the radial, or the cob-web, or any other unusual pattern, should be recorded separately. Later in the laboratory, the readings should be plotted separately. The radial joints have been reported by Hegde, G.V. (1984) and Pujar (1989), while a cob-web pattern of joints, has been reported by Pujar (1989). The cob-web pattern of joints has been reproduced in Fig. 24 and Photo 10.

The columnar joints deserve a special mention. These are hexagonal in shape, and the columns are remarkably long. The columnar joints of St. Mary's Island, Malpe, Mangalore, Karnataka state, constitute a National Monument. These joints are generally vertical in attitude, but inclined ones also have been reported by Gokhale (1996). Hegde, V.N. (1984) has reported horizontal columnar joints in a dyke traversing the granitic rocks of Gangavati-Kamalpuram area, Bellary district, Karnataka state. Mini-columnar joints are sometimes met with, as reported by Patil et.al. (1984). The main purpose of elaborating the columnar joints is to note that this structure is developed only in the basalts, which rocks could be in the form of a flow, or a dyke. It is also a fact that though the Deccan Traps (basaltic flows) are extensively developed, the columnar joints are encountered at very few places. The occurrence of the columnar joints within the dykes, is a rare phenomenon indeed. Therefore during the mapping of the basaltic flows, or the dykes, it should be endeavored not to miss the said joints (if these be actually encountered in the area taken up for the mapping exercise).

SHEARS

These fractures are nearly parallel to each other, are closely spaced, and are numerous in number. The shears are also required to be recorded, which consists of measuring their trend directions. Many times "shear zones" are met with. Hegde, G.V. (1984) has reported 21 shear zones over a span of a mere 32 km. from the quartzarenitic rocks of the southern margin of the southeastern part of the Kaladgi basin, Karnataka state. The shears are characteristically longer than the other kinds of fractures (like the joints) encountered in the rocks. Another feature of distinction, and of importance is that "inbetween two shear zones, the rocks are not at all shattered". This behavior has a bearing on the nature of the deformative forces that had acted on the crustal rocks.

FAULTS

In the case of the faults, its trend, the direction and the amount of inclination, displacement (if any), development of pseudotachylyte, development of smooth and glassy surface (called as the slickensided surface), fine furrows (called as the striations and grooves formed on the fault plane) and gouge, are to be carefully looked for. It is not necessary that all the said features should be developed at each and every fault. The fault plane being a plane of weakness, erosive activity of the geological agents, is generally very strong, along such planes. As a result of the erosion, valleys that are steep and narrow (called as the gorges) are formed. Therefore such valleys should be examined, up and down their course, very closely. The trend of striations, as well as the "direction of smooth feel" when palm of the hand is moved on the "slickensided surface", should be carefully studied and recorded. The trend of the striations, and the direction of "smooth feel", indicate the "direction of the actual movement, along the fault plane". As a matter of fact, the "dextral, the sinistral, the upthrust, the down thrust, the underthrust, the over thrust, the rotational faults" these could be established, only if the direction of movement on the fault plane, is established.

It has been experienced that, instead of displacement (in the sense of the throw of a fault), the movement of one fault-block upon another, causes development of breccia, cataclasite, mylonite, and so on. Therefore, if these features be encountered in the field, the presence of a fault can be surmised at the site of the occurrence of the said features. In the case of rocks having a homogeneous composition (like a granite, flows of basalt), faults could be established only on the basis of the occurrence of the said features. This is so because, eventhough faulting might have taken place, displacement in the form of a throw, is not produced, owing to the absence of any "key horizon" in the rocks, having a homogeneous composition. As an illustration, a slickensided surface reported by Muralidharan (1991), is cited. A very coarse grained, gray granite has been observed by him to have developed quite a prominent and extensive slickensided surface and pseudotachylyte (Photo 11a). In the absence of these features, existence of a fault at that place, could not have been established, because displacement is not at all observable, because the rock is having a homogeneous composition. Desai (1989) has likewise established existence of a fault in the banded hematite quartzites, solely on the basis of the development of bluish green coloured pseudotachylyte (Photo 11b).

Photo 11a. Excellently developed slickenside and pseudotachylyte due to faulting. The significance is that the said features are produced in an extremely coarse grained, porphyroblastic pink granite. Note that the rock has become fine grained along the fault plane, while it is coarse grained away from it. The zone of pseudotachylyte is greenish black in colour, and it is situated inbetween the fine grained portion (extreme right hand side of the exposure), and the coarse grained rock (left hand side of the exposure). The fault plane has an inclination of about 70 degrees. Locality of exposure-2.5 km. SW of Akutothapalli village, Mehboobnagar district, Andhra Pradesh (Courtesy, Dr. D.Muralidharan).

Photo 11b. Typical bluish green coloured pseudotachylyte developed in an exposure of banded hematite quartzites, striking N55°W-S55°E and dipping 65° due S35°W. In the absence of any displacement, the existence of the fault at that place, would have been missed, but for the unmistakable presence of pseudotachylyte. This exposure is located on the northwestern boundary of the Aarpee mines, Jambunath hills. Bellary district, Karnataka state. (Courtesy Dr. H.D. Desai).

SUMMARY

Structural features that are frequently encountered with the igneous the sedimentary, and the metamorphic rocks, have been separately described. Structures common to the rocks irrespective of the category to which they belong, also have been noted. As has been proclaimed, the purpose of this book is to familiarise the field geologists, with the structural features that are often met with, in the crustal rocks. With experience, and continued field work, the geologists can then unravel unusual structural features, too.

CHAPTER 7

Data to be Collected During Mapping Exercise

INTRODUCTION

In this chapter it is desired to deal with the exact data to be collected during the field work (mapping). It is necessary to know the details of the data to be obtained, because, it differs from rock to rock (igneous, sedimentary, and metamorphic), and from structures to structures (joints, folds, faults, lineation, and so on). Even the manner in which the samples of the rocks are to be collected, varies. It is sort of a guide line, in respect of "what is to be observed and recorded" during the field work, that has been described in this chapter. The commoner, yet the essential features have been presented in the following paragraphs.

COLLECTION OF SPECIMENS OF ROCKS

During the field mapping programme, only "insitu" samples of rocks should be collected. The technical term "insitu" means that the rock must exist at that place, at depth, too. It means that, it must not be a "roled" piece of rock. The rivers and the streams bring in numerous pieces of rocks, and many times these are found on the bank, as well in the bed of the rivers and the streams. But these are not "true outcrops" of rocks, and therefore such pieces of rocks should not be picked up as "samples" of rocks. Very close examination is required to confirm the "insitu" character of the "outcrops of rocks".

It is often said that during the field mapping exercise, a "representative sample" should be collected. But it is difficult to categorise any sample to be a "representative one", especially during the course of mapping. Representative sample is a "biassed manner" of collecting any specimen of rock, if the adjective "representative" were to stand for the petrographical character of the specimen. The main purpose of collecting the samples, is to derive the mineralogical composition of the crustal rocks, as accurately as possible. There is yet another feature of importance that deserves a mention. The composition of the rocks, in the context of quality and quantity of the minerals, varies gradually. This is especially true with the igneous rocks. It is also not feasible (and practical) to remove the entire body of the rock, and bring it as a sample to the laboratory, for the further investigations. Therefore a "via media approach" is resorted to. The samples are generally collected in two ways, namely, i) at random, and ii) at regular intervals. In the first method, no specific distance is maintained between the several spots from which samples are collected. In the second method, the samples are collected at "predetermined intervals". This is spoken off as the "Grid pattern" of samples. These two methods of collection of

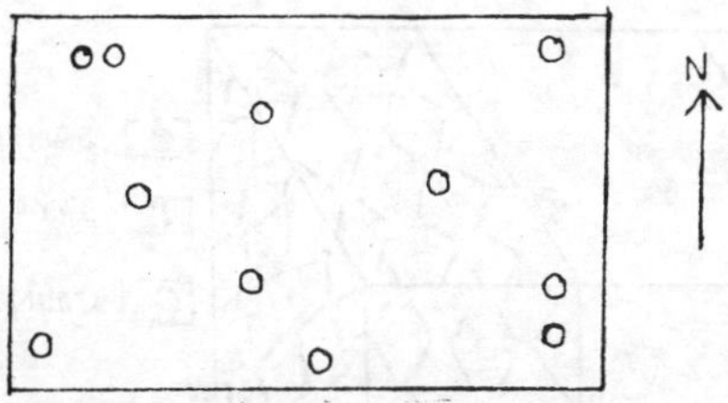

Fig. 25a. Random method of collection of samples of rocks. Circles mark the sites of collection of samples of rocks. Note that the distance between the sample-sites is extremely irregular.

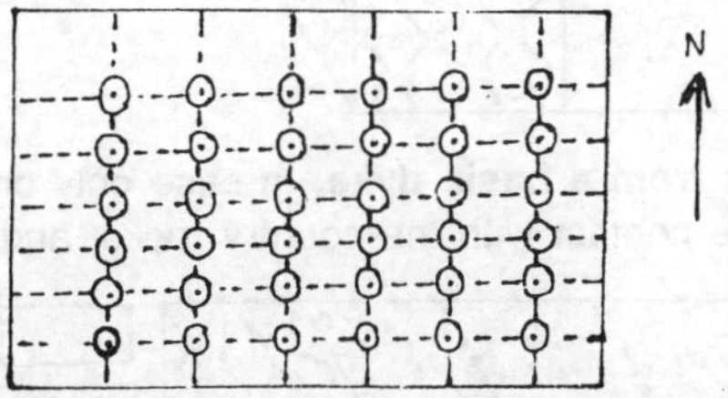

Fig. 25b. Grid method of collection of samples of rocks. Circles mark the sites of collection of samples of rocks. Note that the samples are collected at the intersection of the grids (the dashed lines) and thus maintaining regular interval between the sample-sites.

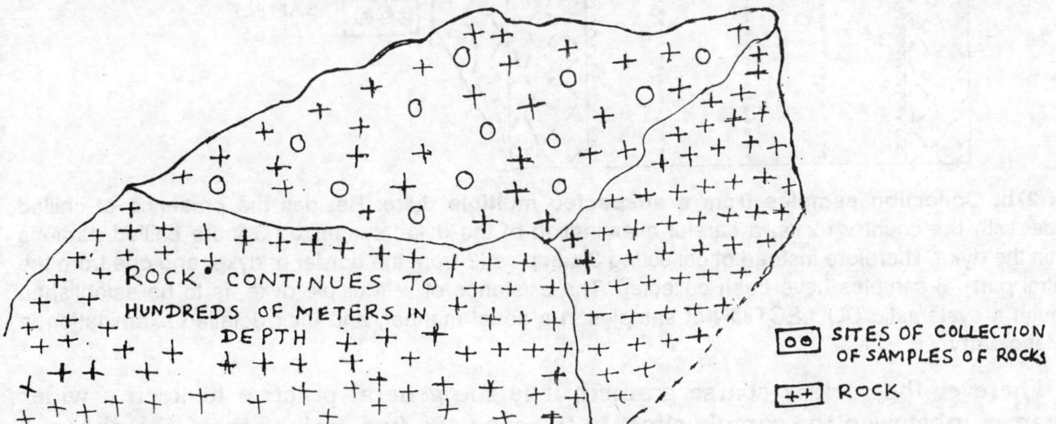

Fig. 26. Depth of surficial samples. Note that the depth of sample is too meagre in comparison with the extent (depth) to which the body of rock continues downwards. Therefore there is need for collection of as many surficial samples as possible, in order to ascertain the mineralogical composition of the rock, fairly accurately. Circles mark the sites of collection of sample, plus marks indicate rocks.

samples, have been shown in Figs. 25 a,b. Whether it is the "random, or the grid pattern", only surficial samples are collected.

In order to visualize the importance of collecting samples of rocks in a systematic manner, Fig. 26 has been drawn. From this figure it is quite apparent that, the depth from which any sample is collected, is "literally skin deep" as compared to the depth upto which the rocks actually extend. Therefore, in order to compensate for the "meagre depth of surficial sample", at least the samples should be collected in a systematic manner, and not on the consideration of them being "representative" ones.

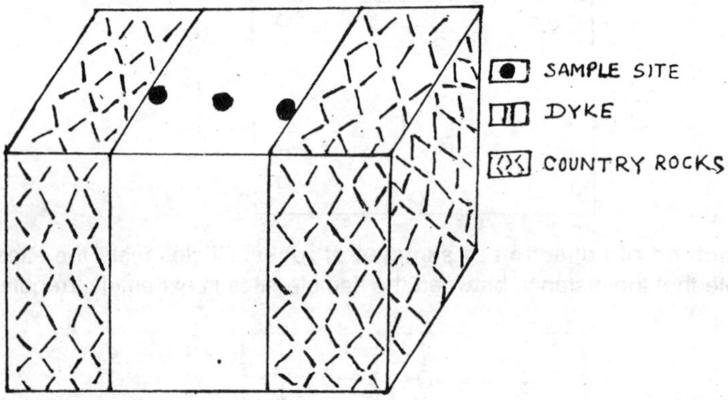

Fig. 27a. Collection of samples from a basic dyke. In case only one basic dyke be suspected, 3 samples are collected: 2 from the contact with the country rocks, and 1 from the central part of the dyke.

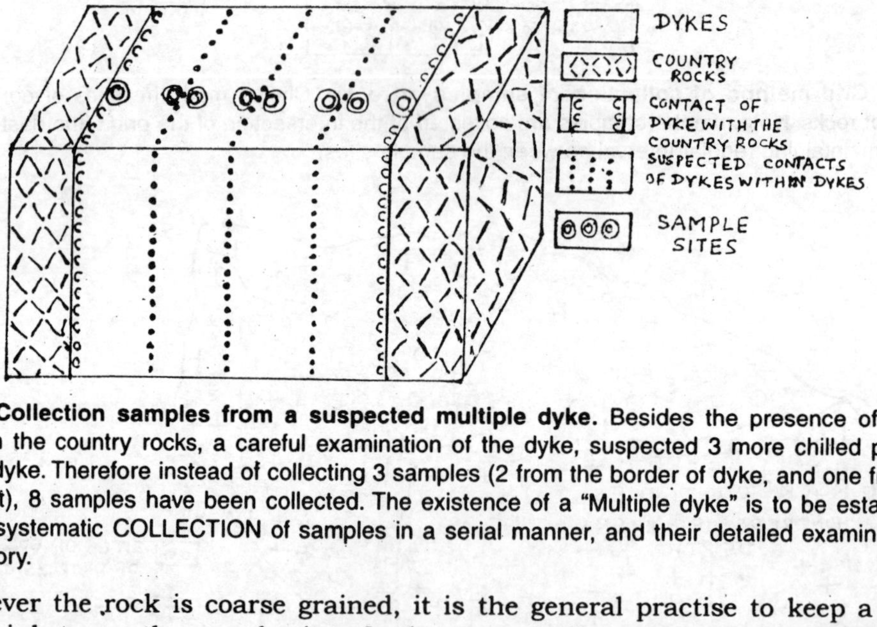

Fig. 27b. Collection samples from a suspected multiple dyke. Besides the presence of chilled border with the country rocks, a careful examination of the dyke, suspected 3 more chilled portions within the dyke. Therefore instead of collecting 3 samples (2 from the border of dyke, and one from the central part), 8 samples have been collected. The existence of a "Multiple dyke" is to be established through a systematic COLLECTION of samples in a serial manner, and their detailed examination in the laboratory.

Wherever the rock is coarse grained, it is the general practise to keep a wider distance, inbetween the sample-sites. In the case of a fine grained rock, the distance between the sample-sites, is to be reduced. The reasoning behind adopting such a procedure is that, in the case of a coarse grained rock, its composition does not change rapidly, and hence a wider distance could be maintained between the sample sites. On the other hand, there is every likelihood of the composition to change rather considerably, if the rock be a fine grained one. However, no specific distance inbetween the sample-sites, could be stipulated. It is therefore said that "more the number of unbiassed samples, the better it helps to ascertain the petrographical (mineral composition) variations of the crustal rocks".

In the case of a basic dyke (dolerite, basalt), a sample each should be collected from its two contacts with the country rocks, and one from its central part. If a multiple intrusion be suspected, samples should be collected from the contact portion of each intrusion (dyke) with the other. These features of collecting the samples from the dykes, have been shown in Figs. 27a, b. In this book, since the ore bodies have bot been dealt with, the methodology of collecting the bore-hole samples has not been described.

COLLECTION OF PETROGRAPHIC DATA

IGNEOUS ROCKS

If the intruded rocks (called as the country rocks) be exposed, the attitude of the surface of contact of the igneous body with the country rocks, i.e., the direction and the amount of inclination, should be recorded. The nature of the contact of the igneous body with the intruded rocks, also should be necessarily recorded. The contact could be sharp or gradational one. The granites (these are the most extensive amongst the igneous rocks) many times contain xenoliths. The shape and the size of the xenoliths should be recorded, but these can not be individually marked on the base map, because their sizes are too small (even if the scale of the base map be enlarged one). However it is advisable to draw separate sketches of the xenoliths, on a convenient scale. If the xenoliths be elongated ones, the direction of elongation should be recorded. If the xenoliths possess structural elements, like the schistocity, the folds, or any other feature, then the trend of schistocity, the attitude of the folds, etc., should be accurately recorded.

The xenoliths generally are the pieces of rocks acquired from the intruded country rocks. The xenoliths are categorised into the i) accidental and the ii) cognate varieties. The accidental xenoliths probably are the result of a forceful intrusion of the magma, into the pre-existing crustal rocks. The cognate xenoliths on the other hand, are produced during the process of crystallisation of the magma that had quietly intruded into the country rocks. Therefore the xenoliths (if encountered) should be very carefully and critically studied. The xenoliths are also described by the other technical terms, namely, the rafts, the inclusions, and the schlierens. There are differences between these terms, but all of them are covered under the general term "xenoliths".

As already noted in the previous chapter, the xenoliths provide important petrographical data, to ascertain the origin of the granitic rocks. The margins of the xenoliths should be carefully studied. The margins could be sharp, or gradational ones. In the latter case, the xenolith will be observed to gradually merge into the granite (granite has been mentioned because this rock contains xenoliths more often than the other igneous rocks) surrounding it. The sharp or the gradational type, are again very useful in deriving the mode of formation of the granites. The xenoliths are logically expected to occur in the vicinity of the intruded country rocks, if these be "accidental xenoliths", while those are generally absent in the core portion of intrusive magmatic body. Therefore the area where the xenoliths are met with, such areas should be accurately marked on the map.

The porphyritic and the porphyry textures are often met with the granitic rocks, and these textures are easily and readily recognisable in the field, even without the assistance of the pocket lens. It is necessary to know the difference between these two textures. In a porphyritic rock, the phenocrysts are few, but are large in size, and these are set in a fine grained matrix (Fig. 28 a, and Photo 12). In a porphyry rock, the phenocrysts are numerous, nearly equal in size, but are small in size, and these are set in a finer grained matrix (Fig. 28 b, and Photo 13). Thus it is apparent that the matrix is more in a porphyritic rock, whereas it is considerably less in a porphyry rock.

As soon as a sample is collected, a detailed description of it, covering its colour, mineral composition (if determinable), texture, unusual features (if any) should be

Photo 12: Exposure of a pink coloured, megaporphyritic granite, at a distance of 500 metres, west of Akutothapalli village, Mehboobnagar district, Andhra Pradesh. Note that the phenocrysts are few, but are big in size, and the groundmass is more in extent. The phenocrysts are of potash felspar, and these are parallely oriented, which is indicated by the handle of the hammer, and the pens. (Courtesy Dr. D. Muralidharan).

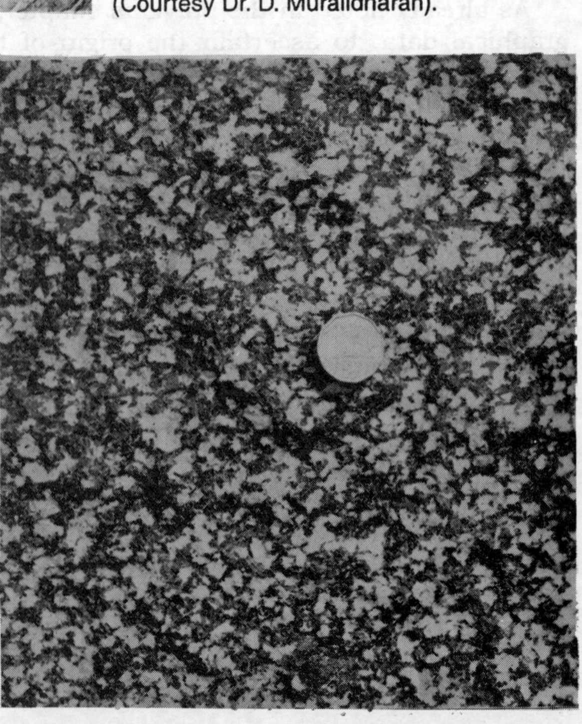

Photo 13: Exposure of a diorite porphyry, at a distance of about 2 km NNE of Lingsugur, Raichur district, Karnataka state. Note that the phenocrysts of felspar are numerous, but are smaller in size as compared to those occurring in Photo 12. Also note that the matrix for the phenocrysts is considerably small in extent, as compared to the space occupied by the phenocrysts (Courtesy, Dr. G.V. Hegde).

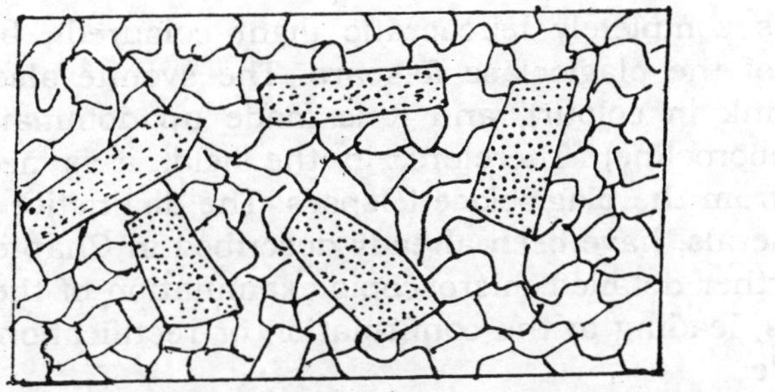

Fig. 28a. Porphyritic texture. Note that the phenocrysts are large, euhedral but few in number, and these are set in a finer grained matrix which is made up of Subhedral to anhedral grains. It is a "Mega porphyritic texture".

Fig. 28b. Porphyry texture. Note that the phenocrysts are numerous, nearly equal in size, but are smaller in size. Also note that the area occupied by the matrix, is considerably small. The grains of the matrix minerals, are anhedral. It is a "Mega porphyry texture".

written immediately. This procedure ensures the petrographical description being correctly noted, against the concerned sample, and not mixing it up with some other sample previously collected. It is also necessary to number the sample immediately. After breaking a piece of rock from the outcrop, it should be trimmed, and as far as possible, a rectangular shape should be given to the smaple. During the process of trimming, chips fall off. A large and a flat chip should be selected, and it should be wrapped up with the specimen in a piece of paper. The chip of rock so collected, is useful in preparing a thin section from it, later in the laboratory.

The lamprophyric dykes possess typical "pitch black" colour. Therefore, a dyke displaying such a colur, may be suspected to be that of a lamprophyre. This rock also possesses porphyritic texture, with ferromagnesian minerals (augite, hornblende, biotite) forming the phenocrysts. Lamprophyre is inherently dark in colour, and added to it, the phenocrysts are also dark in colour. As a result, the porphyritic texture (which is also called as the panidiomorphic texture), is not easily detectable, during the field mapping exercise. However, an attempt should be made to observe the said petrographical features, in order to locate the lamprophyric dykes in the field. These dykes are not frequently met with, and hence a special emphasis is required to be laid, to discover them.

The monomineralic rocks like the dunite, the pyroxenite, the peridotite, the picrite, the hornblendite, are all dark in colour. The other diagnostic feature is the possession of a high density. The specimens of these rocks are therefore heavy, as compared to the other igneous rocks. Therefore, a combination of a dark colour, and a high density, enables to distinguish the said monomineralic rocks in the field. There is yet one more monomineralic rock, which is called as "Anorthosite". This rock however, is completely leucocratic (light coloured), because it is made up almost entirely of the plagioclase felspars. The syenite also is a leucocraitc rock (actually it is pink in colour), and it is made up dominantly of potash felspars (orthoclase or microcline). Therefore, in the field, it is necessary to distinguish potash felspars from the plagioclase felspars. The distinctive megascopic characters of these two minerals, have been already described in Chapter 2. Such a step helps to undertake further detailed microscopic examination of the thin sections of such leucocratic rocks, leading to the confirmation or rectification of the field identification of the sample.

It is considerably difficult to identify the volcanic rocks in the field, on the basis of the megascopic petrographical features, because these rocks are extremely fine grained to glassy in nature. The *basalt* alone possesses the amygdaloidal structure, but it does not mean that every basalt must develop the said structure. The *trachyte* is to be identified from its characteristic flow structure, s much so that the structure itself is named as the "trachytic structure". The *pumice* alone possesses the scoriaceous structure. Owing to this, the surface of a pumice is very rough and one gets a feeling of "pricks", when palm of the hand is run over the rock. The *obsidian* displays a smoky colour and a perfect conchoidal fracture. The *rhyolite* displays "colour banding" and it is leucocratic unlike the other volcanic rocks. Therefore these aforesaid features should be made use of, in distinguishing the volcanic rocks. However, the microscopic examination of the thin sections of these rocks, is absolutely necessary to make a specific and a final identification of the samples of the volcanic rocks.

SEDIMENTARY ROCKS

These rocks are identified more on the textural, than on the petrographical (mineral composition) characters. The sandstones and the limestones almost look alike, because both are light in colour. Therefore, a treatment with dil. HCl should be carried out on the samples, besides ascertaining the presence of the grains of quartz in the specimen. The limestone alone gives effervescence on treatment with dil. HCl acid, and it does not contain the grains of quartz in it. On weathering, the limestones develop a buff colouration on the surface. Further, a sample of limestone could be easily scratched with a penknife, whereas a sample of sandstone is hard to scratch. If the grains of quartz be noticed in the sample, that indicates the sample to be of a sandstone, and not of a limestone.

In the case of a conglomerate, the dimensions of the pebbles and the boulders occurring in it, should be determined. If too big, the rock is called as a boulder bed. If the pebbles and the boulders be of a medium size, the rock is classified as a conglomerate. If the pebbles be fine in size, the rock is named as a grit. If the pebbles be angular, and irregular in shapes and sizes, the rock is classified as a breccia. The shales are very fine grained rocks (even finer than a fine grained sandstone) and possess different colours like red, buff, white, purple, black, violet. Extremely find bedding planes are noticed in them. These rocks possess cleavage, it being called as the "fissility". Thus with the combination of a fine grain size, cleavage, and composition, the shales are distinguished from the sandstones.

It has been already noted that the rock graywacke is by and large, a sandstone, but its colour is gray. The graywacke is somewhat cold in its feel. Therefore in a sedimentary terraine, if gray coloured rocks be encountered, those should be examined more closely and carefully, with the prospect of recording graywackes from the terraine taken up for mapping.

METAMORPHIC ROCKS

The most salient feature of these rocks is the development of cleavage, and the orientation of the minerals. These two features together give rise to the schistocity, the foliation and the lineation. Therefore these features are to be recorded during the field work. A sandstone and a quartzite differ mainly in respect of the texture. A sandstone is made up of rounded to sub-rounded grains of quartz that are set in a matrix. In a quartzite, the grains of quartz posses "serrated or sutured borders" and there is a total absence of any matrix for the grains of quartz. These two features of the quartzites could be easily observed by the use of the pocket lens. Therefore the said data (features) should be necessarily collected during the field mapping exercise. Likewise, the development of the slaty cleavage (in a slate), the schistocity (in a schist), the foliation (in a gneiss), should be recorded. The salient metamorphic minerals like the garnet, the kyanite, the sillimanite, the staurolite etc., should be identified. The augen structure is more commonly encountered in the metamorphic rocks like the granite gneiss, the quartzites and so on. Some of the slates display a spotted appearance. The spots consist of minerals (which have just commenced the process of recrystallisation-called as the incipient grains) and such a rock is called as a "spotted slate". As said earlier, the metamorphic rocks, share the petrographic features of the igneous and the sedimentary rocks, and such features should be necessarily observed and recorded, because, these help to infer the parental rocks (whether sedimentary or igneous) that had undergone metamorphism.

COLLECTION OF STRUCTURAL DATA

INTRODUCTION

The structures produced in the rocks are either planar or linear ones. It means that in most of the cases, the planes (surfaces) are produced like the joint planes, the bedding planes, the fault planes and so on. In the "linear category", orientation of minerals devoid of any planes, is developed. The data to be collected in respect of the two categories of structures are described below.

PLANAR STRUCTURES

The direction and the amount of inclination (dip) of the planes should be recorded. In the case of joints, wherever quarries are available, numerous readings of joints should be taken. It is enough to record the direction of dip, because from this direction, the direction of strike can be calculated, but vice versa is not true. In order to construct the stereograms, a minimum of 200 readings of dip direction and the amount of dip of the joint planes, are necessary. Such readings are also useful in constructing the histograms and the rosette diagrams. The kind of the rock in which the joints are developed should be noted. Separate readings should be taken for the different quarries (even if the rock might be same), and also for the different rocks. Separate stereograms should be constructed for each quarry, or rock, or spot of collection of data, later in the laboratory. The procedure to be adopted in the construction of the stereograms has been described by Gokhale (1996). The joint planes sometimes produce specific patterns like the columnar, the rhombic, the rectangular and so on. Field photogrpahs of such instances should be necessarily taken. Any unusual pattern also should be photographed. Gokhale et.al. (1983) have reported "brick work" like joints developed in the banded hematite quartzites, Nagavi village, Gadag district, Karanataka state (Photo 14).

Photo 14: Brick-work like joints developed in the banded hematite quartzites exposed in the SE vicinity of Nagavi village, Gadag district, Karnataka state. Length of brick (the joint block) is parallel to the strike direction of the rocks. The outcrop of this peculiar "brick-work like joint pattern", is about 25 sq. meters in extent. It is located at a distance of 1.5 km. S25°E of Nagavi village. Note the perfect rectangular (brick like) shape of the joint blocks. (Courtesy Dr. S.C. Puranik).

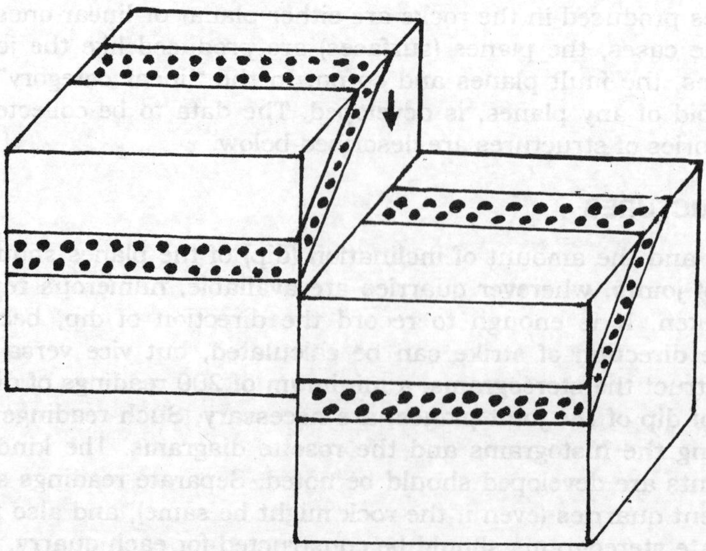

Fig. 29 a. Fault of translation. Note that the amount of dip of the bed (dotted part) has remained same in the two blocks, even after faulting.

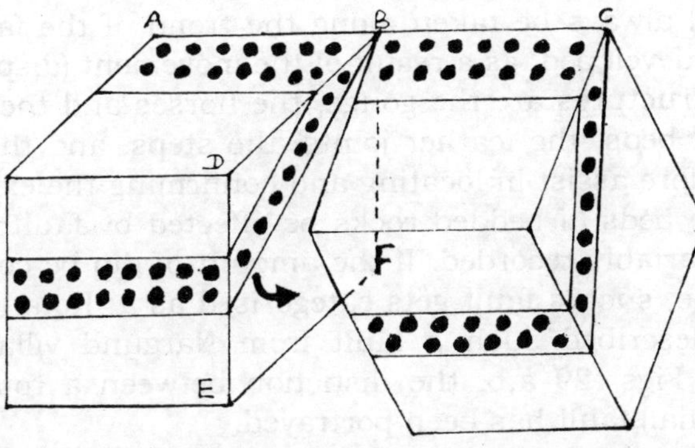

Fig. 29 b. A rotational fault. Note that the amount of dip of the bed (dotted part) in the "right handside block" has increased, because it has undergone rotational movement along the fault plane BDEF. The movement is indicated by a "curved arrow, unlike a straight one as shown in Fig. 29 a. ABC marks the "hinge" of the fault, and such faults are therefore called as the "Hinge Faults". The point of importance is that the "rotational movement" is required to be established by recording the amount of dip across suspected portion of outcrop, like the line BD in this figure, which emerges as the trace of a rotational fault.

In the case of the fault planes, whether it is a "planar or a curviplanar surface", its attitude should be accurately recorded. The striations are many times produced on such planes. Their (striations) trend directions must be recorded. Sometimes 2 or 3 sets of striations are developed. In such cases, their relative ages (meaning, which was formed first, and which was formed later) should be established in the field. Pujar (1989) has reported one such instance from the vicinity of Basidoni village, Saundatti taluka, Belgaum district, Karnataka state. He has observed two sets of striations; one set trending in a N30°W—S30°E direction, the other set trending in a N70°E—S70°W direction. This latter set has been proved to be younger, because, these striations are continuous, whereas, those trending in a N30°W—S30°E are obscure and patchy. The reasoning put forth in arriving at such a conclusion is as follows. On the fault plane, the movement took place in a N30°W—S30°E direction, and this resulted in striations also trending in a similar direction. The fault plane then witnessed another movement in a N70°E—S70°W direction, which went on wiping out the previously developed striations trending in a N30°W—S30°E directions. The striations trending in a N30°W—S30°E directions therefore became patchy and discontinuous ones.

The striations indicate the actual direction of movement of rock, along the fault plane. Therefore, wherever striations are noticed, those should be necessarily recorded. Such a data along will help to establish faults such as the dextral, the sinistral, the upthrust, the strike-slip fault, the oblique-slip fault and so on.

Many times faults are suspected through the presence of pseudotachylyte, cataclasite, gouge and so on. Therefore, the said features play an important role in the recognition of the faults. Many times, the fault planes are mineralised, quartz constituting the most frequently occurring mineral. Such a mineralisation also helps to locate a "concealed fault', especially in rocks having a homogeneous composition. In such rocks, the actual displacement is not detectable eventhough, faulting might have taken place.

A traverse should always be taken along the trend of the fault, because, many other structures are developed, as a result of the movement (displacement) along the fault plane. These structures are the gouge, the horses and the slices, the breccia-tion, the dragging of beds, the feather joints, the steps, and the fracture cleavage. These features therefore assist in locating and confirming the existence of fault. The amount of dip of the beds (if bedded rocks be affected by faulting) across the fault plane, should be invariably recorded. If the amount of dip be considerably different across the fault plane, such a fault gets categorised as a "Rotational fault". Puranik et.al. (1982) have described such a fault from Nargund village, Gadag district, Karnataka state. In Figs. 29 a,b, the distiction between a rotational and a non-rotational (translational) fault has been portrayed.

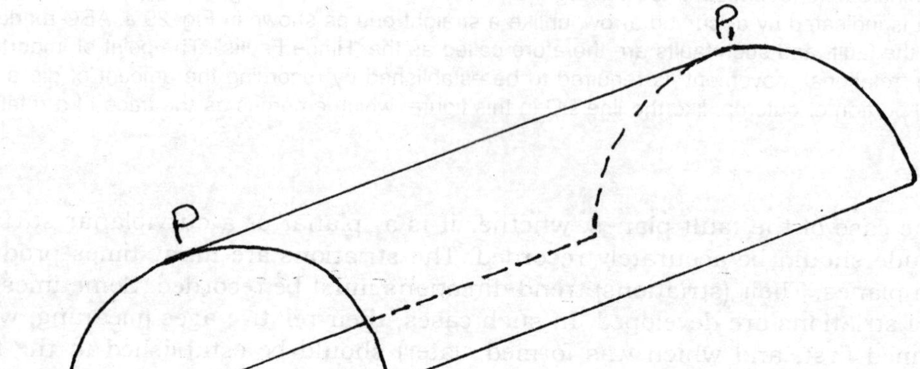

Fig. 30a. A non-plunging fold. Note that axis PP1 of the anticline is horizontal.

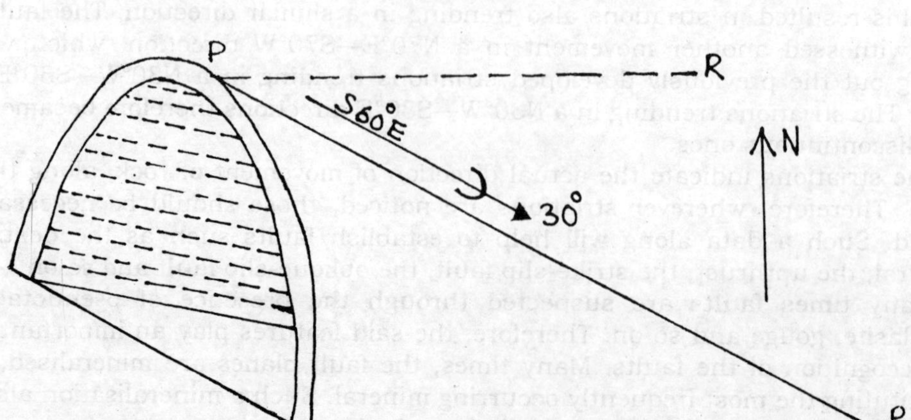

Fig. 30b. A plunging anticline. Note the axis PP1 of the anticline is inclined, the amount being 30 degrees. PR is horizontal line, and angle RPP1 is 30 degrees. PP1 is the direction of plunge which is due S60E. (After Fig. 55a of Koppad, 1976).

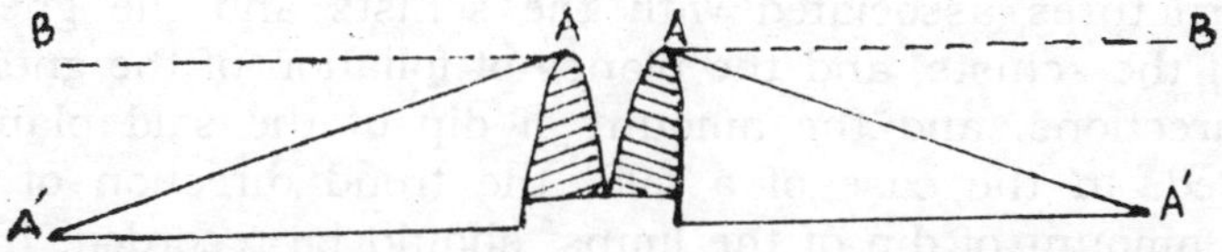

Fig. 31a. A doubly plunging fold. Note that the axis AA′ of the anticline dips in opposite directions. Angle BAA′ is the plunge of the anticline.

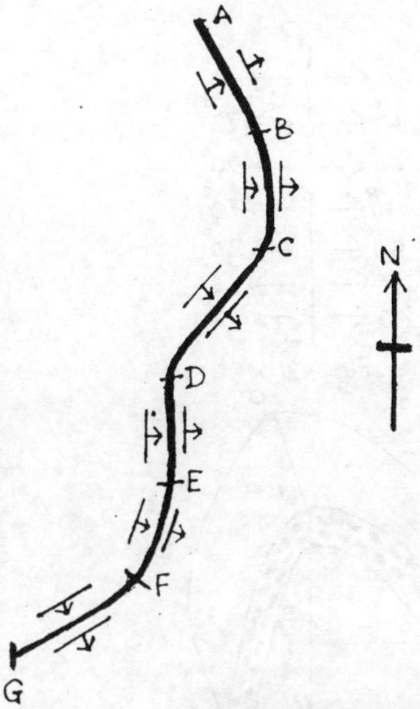

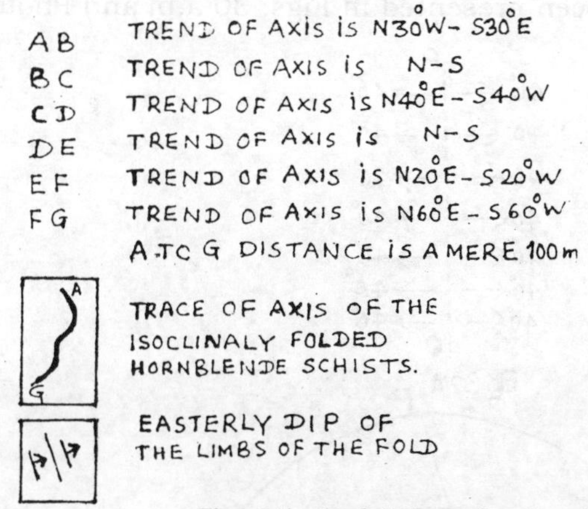

Fig. 31b. Refolding of axis of a fold. Fig. 31b is drawn on the data set out by Koppad (1976). The feature to be stressed is that the refolded nature of the isoclinal fold could be deciphered through recording and plotting the direction of dip of schistose planes.

Photo 15: An exposure of a plunging anticline developed in the garnetiferous amphibolites, in Doddahalla stream, Kognur village, Shirhatti taluka, Gadag district, Karnataka state. The fold plunges due south i.e., away from the observer. A geologist is standing right on the axis of the plunging fold. The divergent directions of dip (due east and west) of the anticline can be clearly observed. (Courtesy Dr. V.B. Koppad).

The planar structures associated with the schists and the gneisses are the cleavage planes of the schists, and the planes of foliation of the gneisses. The dip and the strike directions, and the amount of dip of the said planar structures should be recorded. In the case of a. fold, the trend direction of the axis, the direction and the amount of dip of the limbs, should be recorded. If the fold be a plunging one, then the direction of the axis and its inclination (which are called as the bearing and the plunge, respectively), should be recorded. These features of the folds have been presented in Figs. 30 a,b, and Photo 15.

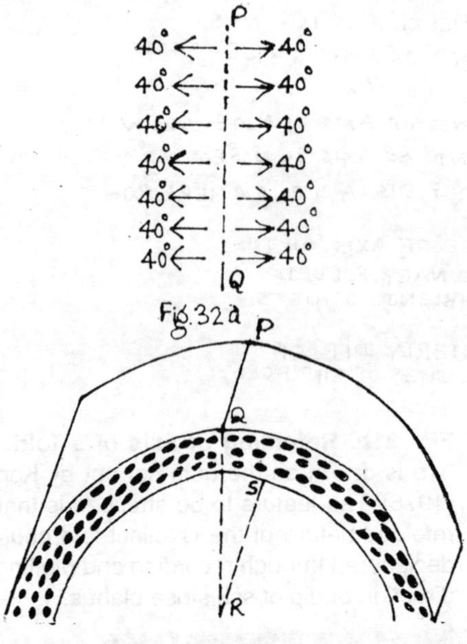

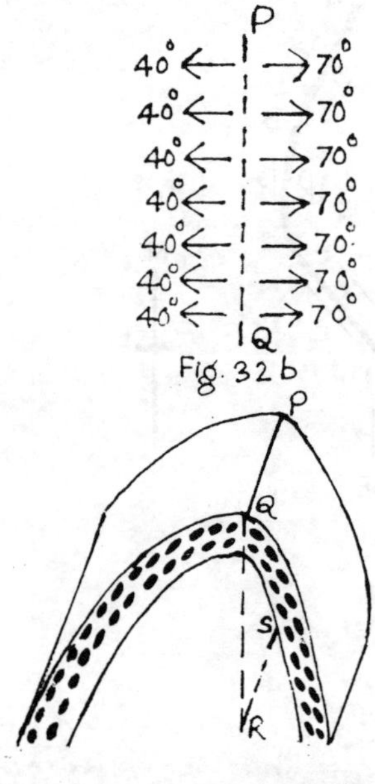

Fig. 32a. represents field data of a folded structure. Note that the amount of dip of the two limbs is the same. The dashed line marks the axis of the fold. Fig. 32a, is derived from Fig. 32a. The symmetrical aspect of the anticlinal fold can be easily appreciated in Fig. 32a,. PQRS is the axial plane of the fold, while PQ is its axis.

Fig. 32b. represents field data of a folded structure. Note that the amount of dip of the two limbs is unequal. The dashed line marks the axis of the fold. Fig. 32b, is derived from Fig. 32b. The asymmetrical aspect of the anticlinal fold can be appreciated better in Fig. 32b,. PQRS is the axial plane of the fold, while PQ is its axis.

A fold could be doubly plunging one. In such cases, the direction and the amount of plunge of the two axes, should be recorded, besides recording the direction and the amount of dip of the limbs of the fold. A doubly plunging fold has been shown in Fig. 31 a. At times, even the axis of the fold gets refolded, which results in the formation of a "refolded fold". Koppad (1976) has reported such a fold, and it is presented in Fig. 31 b.

The amount of dip of the two limbs of the fold must be necessarily recorded, because, only then the symmetrical and the asymmetrical folds could be established. This has been shown in the form of field data in figs. 32 a,b, and the folds derived therefrom in Figs. 32 a_1 b_1.

The point to be stressed is that wherever planar structure is encountered, its inclination and the direction, should be recorded with meticulous care and accuracy.

LINEAR STRUCTURES

These structures are not readily detectable during the field mapping exercise, unless the outcrops of rocks are carefully and very minutely scanned. To the category of the linear structures, belong the shears, the longer axes of the grains (minerals or pebbles), the intersection of the bedding planes and the planes of schistocity, the striations on the fault planes, ridges of the oscillation and the current ripples, streaks of the ferromagnesian minerals (mostly biotite in granites) and so on. The trend direction of the said linear elements, should be carefully recorded. In the lath shaped and the accucular minerals (like the felspars and the amphiboles) producing the linear structure, the inclination (called the plunge) of the minerals, in the planes of schistocity or those of foliation, should be measured. This liner structure which is described as "lineation" is invariably associated with the schists and the gneisses. The trend (called as the bearing) of the longer axes of the minerals also should be recorded.

DATA FOR PETROFABRIC STUDIES

This is a specilaised and an important aspect of structural studies. Many times it is desired to ascertain the direction in which the deformative forces had acted on the rocks, resulting in the development of several structures. Due to these forces, externally visible structures like the joints, the folds, the faults and other structures are produced. It is also logical to expect, some internal changes to have taken place within the framework of the minerals composing the rocks. The deformative forces also could affect the constituent minerals, and bring about optical changes within them. Such deformative forces affecting the minerals, are called as the "penetrative forces". The "optic axis" of minerals (like quartz) might develop parallelism (orientation), though externally the grains of quartz might appear to be unoriented (meaning lack of parallelism). Likewise, biotite mica, which has only one set of perfect cleavage, the cleavages (which appear as lines, when observed in thin sections of rock) might display parallelism. These structural aspects constitute the subject of the "Petrofabric studies". These optical properties of the minerals are utilised in the petrofabric analysis.

Since the object of the petrofabric studies is to derive the direction of the deformative forces, "oriented samples" of rocks are required to be collected. The petrofabric studies are generally carried out in the terraines, made up of the metamorphic rocks, or where faults are suspected, and so on. The best suited rocks for such a study are the quartzites, and those that contain numerous grains of quarts, and /or plates of biotite in them. A large and a fresh (undecomposed) chunk of rock is broken from the outcrop. The size of the broken sample should be 8×8×8 cm., because, in the laboratory, THREE sections that are mutually perpendicular to each other, are required to be obtained from it. The chunk of rock smaples is placed back at the site from where it was broken. After ensuring that the sample has regained its original disposition in the outcrop, "north line" is drawn on the sample, with the help of a clinometer compass. The other aspects of the laboratory proce- dure concerning the "petrofabric analysis" have been detailed out by Gokhale (1996).

SUMMARY

The general aspects of the collection of data during the field mapping exercise, have been described. As already noted, prior to starting the field studies, data (information) described by the earlier workers, should be collected, which gives a general idea, about the geological features to be expected in the area taken up for the geological mapping. It is suggested that, the petrographical and the structural data, should be plotted, on two separate base maps, in order to avoid complication leading to confusion. Likewise, there should be two sets of base maps; one set to be carried daily to the field in order to collect data, and the second set to be left behind at the "camp site". The data collected in the field, should be transferred eyeryday, onto the maps kept at the "camp site". Such a procedure ensures very little loss of data that was collected in the field. It is also suggested that a separate base map should be reserved, to mark the sites of the collection of the samples of rocks. All said and done, inspite of all these guide lines, *the geologist has to develop the "field vision". The rocks do talk to us, provided we learn their language. The geologist has to proceed for the field work with a mind, and a tremendous patience to undertake dead walk. No stone should be left unturned. Incomplete and an insufficient data, will give an erroneous constitution of the crust of the earth.*

CHAPTER 8

Some Field Indicators of Minerals, Rocks and Structures

INTRODUCTION

So far the various aspects of the geological mapping, have been described. The two components of a geological map, are the rocks, and the structures possessed by them. It is thus apparent that the geologist should be in a position to identify the various rocks in their field habitat, and also to recognise the structures developed in them. The identification of the rocks, is especially required to be based on the megascopically observable features. Of course, such an identification is to be later confirmed or rectified, through the microscopic examination of the thin sections of the rocks, in the laboratory. The structures however, are required to be recognised only in the field, and not in the laboratory. It is therefore thought proper to enumerate certain indicators that will be found useful in the field, to make a preliminary identification of the rocks, the minerals, and the structures. It is desired in this chapter to enlist some of the indicators.

INDICATIONS FROM GEOMORPHOLOGY

INDICATOR	WHAT IT INDICATES
1. Elongated hillocks with steep side slopes, also called as "ridge topography".	1. Hard, resistant rocks possessing high inclination, or nearly vertical-quartzite, banded hematite quartzite, or a basic dyke (dolerite, basalt) having considerable length.
2. Hogback scenery, hillock having medium to high side slopes.	2. Basic dyke (dolerite, basalt) or sedimentary rocks having a moderate to high inclination.
3. Flat topped hills, Mesa or Butte forms, Trap scenery.	3. Bedded sedimentary rocks with almost horizontal attitude, or horizontal lava flows (basalt, rhyolite etc.).
4. Tor scenery, Watermelon topography, balanced rocks.	4. Granitic rocks.
5. Hills with gentle slope on one side, and a steep, almost vertical slope on the opposite side.	5. Dipping sedimentary rocks, the gentle slope marking the dip direction of the beds (rocks).
6. Narrow valley with almost steep sides - called as gorge.	6. Presence of fault trending along the course of the valley/gorge.

7. Unusually straight course of a river/stream, straight sea coast.	7. Presence of a fault along the straight course of river/stream, or sea coast.
8. Waterfall, spring.	8. Presence of fault.
9. Wind gap.	9. Presence of fault.
10. Conical hill.	10. Presence of a centre of volcanic activity and therefore presence of volcanic rocks.
11. Presence of numerous meandering bends in a stream or river.	11. Structural weakness like presence of numerous parallel shears, minor faults.
12. Off set ridges.	12. Faults at the places of off-setting.

INDICATIONS FROM EXTERNAL APPEARANCE OF ROCK

INDICATOR	WHAT IT INDICATES
1. Spheroidal weathering	1. Basic igneous rocks—basalt, dolerite, gabbro.
2. Vesicular structure	2. Volcanic rocks.
3. Amygdaloidal structure	3. Basalt.
4. Scoriaceous structure plus prickly feel on the surface of the rock.	4. Pumice—an acidic igneous, volcanic rock.
5. Smody colour, glassy appearance with a conchoidal fracture.	5. Obsidian—a volcanic rock.
6. Elephant skin weathering.	6. Gabbro.
7. Pillow structure.	7. Spilite—a soda rich, basic volcanic rock.
8. Ropy structure.	8. Basic volcanic rock-basalt.
9. Rock giving a musical ring when hammered.	9. Phonolite—an unsaturated volcanic rock. This rock was previously called as a clinkstone.
10. Extremely dark coloured (pitch black) rock with a panidiomorphic texture (ferromagnesian minerals constituting the phenocrysts).	10. Lamprophyre. It occurs as dykes.

INDICATIONS FROM TEXTURE, COLOUR, STRUCTURE

INDICATOR	WHAT IT INDICATES
1. Light coloured, coarse grained, graphic texture, rapakivi structure.	1. Granite, Granodiorite.
2. Pink coloured coarse grained rock	2. Syenite.
3. Orbicular structure	3. Diorite

4. Light colour, extremely coarse grains of felspars, quartz and biotite.	4. pegmatite. It occurs as dykes and veins.
5. Light colur, saccharoidal (sugary) texture, and made up of felspars and quartz.	5. Aplite. It occurs as dykes and veins.
6. Greenish colour, displays cleavage planes.	6. Slate (if cleavage be perfect, and finer grain size). A schist (constituent minerals identifiable, because of coarser grain size). Possibly, it could be amphibolite.
7. A fine grain size, rock giving copious effervescence when treated with dil. HCl acid.	7. Limestone, Marble.

INDICATIONS FROM MINERAL COMPOSITION

INDICATOR	WHAT IT INDICATES
1. Quartz grains (rounded to subrounded) set in a matrix	1. Sandstone.
2. Quartz grains with sutured borders, and absence of any matrix for the grains.	2. Quartzite.
3. As in 2 above	3. Vein quartz.
4. Quartz+potash felspars+biotite	4. Granite.
5. Quartz+potash felspars+biotite. Rock possessing extremely coarse grain size.	5. Pegmatite. It occurs as dykes and veins, cutting granitic rocks.
6. Quartz+potash felspars, Extremely light coloured (because the rock lacks ferromagnesian minerals)., displays saccharoidal (sugary) texture (called as the aplitic texture).	6. Aplite. It occurs as dykes, veins cutting granitic rocks.
7. Potash felspars predominate, +hornblende, without or with little quartz.	7. Syenite. Possesses typical pink colour. It occurs in the vicinity of granites.
8. Plagiocalse felspars+pyroxenes.. Coarse grain size.	8. Diorite, Gabbro.
9. As in 8 above, but a finer grain size.	9. Dolerite, Basalt. These occur as dykes or as flows.
10. Pyroxenes predominant.	10. Pyroxenite.
11. Quartz grains+clay minerals, Rock loosely compacted.	11. Shale, Argillite, Mudstone.

12. Greenish or gray coloured rock containing chlorite or biotite, and grains of quartz, +any one of the following mineral-garnet, kyanite, staurolite, amphiboles, pyroxenes.	12. A schist. Such rocks display typical schistocity. Depending upon the dominant ferromagnesian mineral, the rock is to be named, as Chlorite schist, Biotite schist, Hornblende schist, Kyanite schist and so on.
13. Felspars, quartz and flakes of biotite, or plates of hornblende. The rock is light in colour, and displays foliation and/or lineation.	13. A gneiss. Dominantly occurring rock is a granite gneiss.

INDICATIONS FOR MINERALS

INDICATOR	WHAT IT INDICATES
1. Colourless, anhedral, cleavages absent.	1. Quartz.
2. Colourless, sometimes pink in colour, two sets of pinacoidal cleavages, opaque or translucent, occurs as laths.	2. Potash felspar (Orthoclase or Microcline).
3. Colourless, occurs as plates which are water clear (transparent)	3. Plagioclase felspars.
4. Prismatic in habit. Two sets of cleavages. Display a bladed form.	4. Amphiboles.
5. More equant grains in comparison with those of the amphiboles. Two sets of cleavages.	5. Pyroxenes
6. Flaky habit. Perfect one set of basal cleavage.	6. Mica. Biotite is black in colur, Muscovite is light in colur.
7. Hard, usually pink in clour. No cleavages. Commonly associated with the schists.	7. Garnet.
8. Typically blue in colur, and a perfect bladed form. Associated with the schists.	8. Kyanikte.
9. Perfect rhomhohedral claeavage. Very soft, and gives effervescence with dil. HCl acid.	9. Calcite.
10. Very soft, and soapy to touch.	10. Talc, Steatite.

SUMMARY

Field indicators of the commonly and extensively encountered crustal rocks, minerals, and structures (faults, dipping beds etc.) have been enumerated. These will help to identify the rocks, the minerals and the structures, with a fair degree of accuracy.

CHAPTER 9

Symbols Used During Geological Mapping Symbols Used on Topographical Maps

Howsoever large a scale be chosen for preparing the geological maps, the space available on the base map, is still very small to accommodate the structural elements like the joints, the folds, the faults etc. Therefore symbols have been designed to indicate some of the structures. Symbols are also made use of to represent petrographical varieties like the granite, sandstone and so on. But those symbols

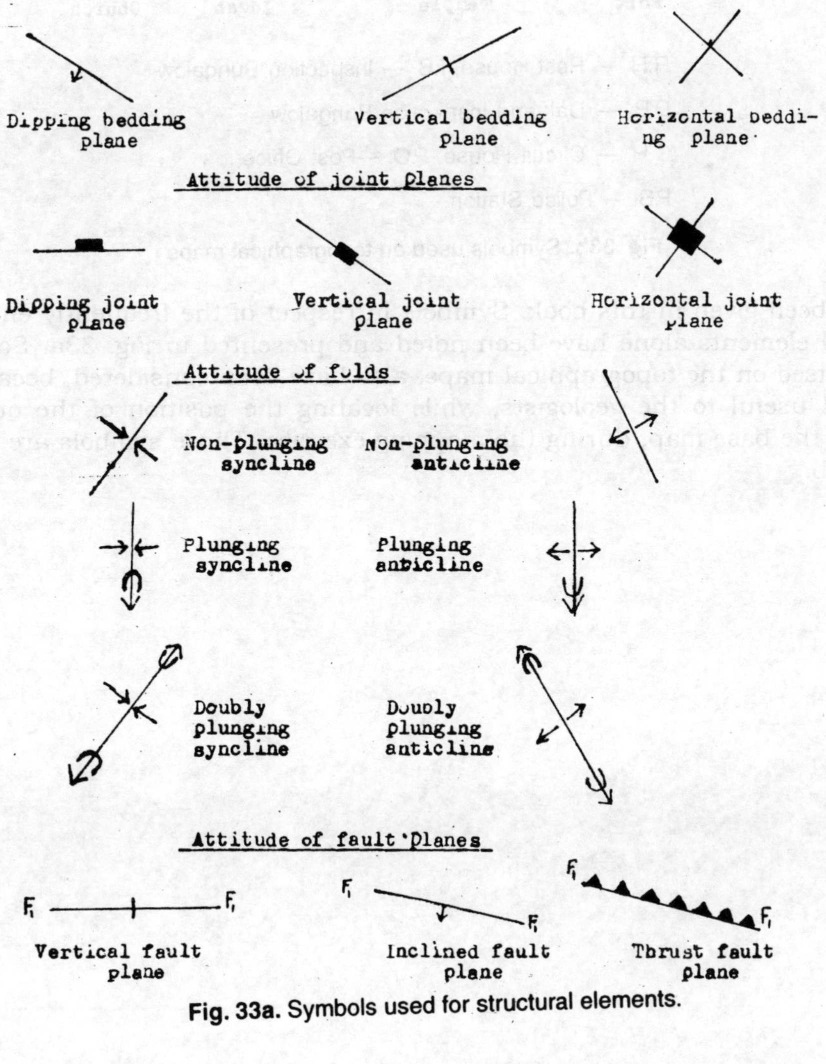

Fig. 33a. Symbols used for structural elements.

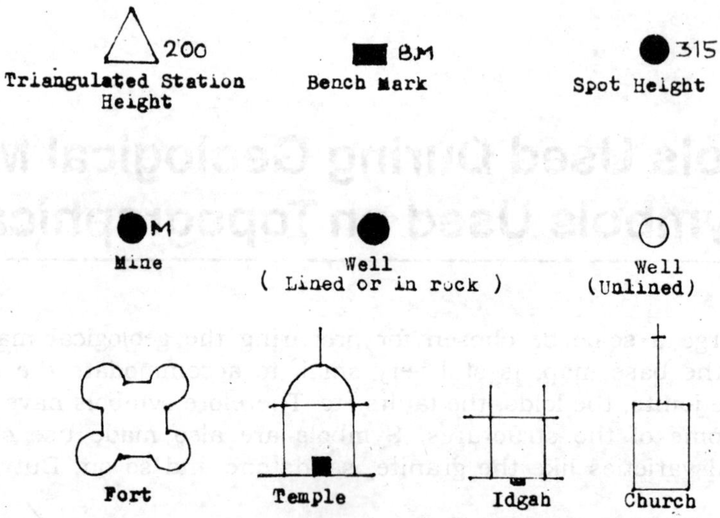

R.H. — Rest House, I.B. — Inspection Bungalow,

D.B. — Dak and Inspection Bungalow,

C.H. — Circuit House, P.O. —Post Office,

P.S. — Police Station.

Fig. 33b. Symbols used on topographical maps

have not been given in this book. Symbols in respect of the frequently encountered structural elements alone have been noted and presented in Fig. 33a. Some of the symbols used on the topographical maps, also have been considered, because those are found useful to the geologists, while locating the position of the outcrops of rocks, on the base map, during the mapping exercise. Those symbols are presented in Fig. 33b.

References Cited

Cloos, H. (1939); Hebung, spalting, vulcanismus. Geol. Rund. Band 30, Zwisschheft, 4a, pp.406-527.

Crompton, R.R. (1968); Manual of field geology, Wiley Eastern Private Limited, New Delhi.

Deendar, D.I. (1982); Geology of the Vengurla area, Ratnagiri district, Maharashtra state. Unpub. Ph. D. thesis, Kar. Univ. Dharwad.

Desai, H.D. (1991); Geological studies of the rocks at and around Jambunath hills, Bellary district, Karnataka state. Unpub. Ph.D. thesis, Kar, Univ. Dharwad.

Gokhale, N.W. (Dec. 1965); Petrological, mineralogical, petrochemical and structural studies of the granites and metamorphic rocks of Velence mountains. Unpub. Ph.D. thesis, Hung. Acad. Sc. Budapest, Hungary.

Gokhale, N.W. (1970); Structural studies on the granites and the associated schists of the Velence mountains, Hungary, and the granite emplacement. Acta Geol. Acad. Sc, Hung. Tomus 14, pp. 5-22.

Gokhale, N.W. and Koppad, V.B. (1971); Xenolith bearing basalt dyke of Belhatti, Shirhatti taluka, Dharwar district, Mysore state. Kar. Univ. Sc. Vol.XVI, pp. 52-58.

Gokhale, N.W. and Puranik, S.C. (1981); On the occurrence of spherulitic structure in basic dyke from Doni area, India. Acta. Geol. Acad. Sc. Hung. Vol. 24 (2-4) pp. 331-336.

Gokhale, N.W., Gourashettar, V.K. and Puranik, S.C. (1983); Peculiar jointing in banded hematite quartzites of Nagavi, Gadag taluka, Dharwad district, Karnataka state, Prof. Kelkar Memorial volume, Ind. Soc. Earth Scientists, pp. 25-28.

Gokhale, N.W., Waghamare,B.P. and Hegde, G.V. (1986); A note on the occurrence of pillow structures in the schistose rocks from Attikatti-Mahalingpur area, Gadag schist Belt, Dharwad district, Karnataka state. Kar. Univ. Jour. Sc. Vol. XXXI, pp. 88-92.

Gokhale, N.W. (1991); A manual of problems in structural geology. CBS Publishers and Distributors, New Delhi.

Gokhale, N.W. (1996); Theory of structural geology. CBS Publishers & Distributors, New Delhi.

Gothe, N.N. (1973); Geology of the granitic and the associated rocks of Mundargi-Hadagalli area, Karnataka state. Unpub. Ph.D. thesis, Kar. Univ. Dharwad.

Hegde, G.V. (1984); Structural and sedimentological studies of the rocks of Mudkavi-Lakhamapur area, Belgaum-Bijapur districts, Karnataka state. Unpub. Ph.D. thesis, Kar. Univ. Dharwad.

Hegde, V.N. (1984); Geology of the Gangavati-Kamalpuram area, Raichur—Bellary districts, Karnataka state. Unpub. Ph.D. Thesis, Kar. Univ. Dhrarwad.

Huang, W.T. (1962); Petrology. McGraw Hill Book Company, PP. 82-85.

Koppad, V.B. (1976); Geology of the area around Belhatti and Bannikoppa, Shirhatti taluka, Dharwad district, Karnataka state. Unpub. Ph. D. thesis, Kar. Univ. Dharwad.

Lahee, F.H. (1987); Field geology. 6th Edition, CBS Publishers & Distributors, New Delhi.

Muralidharan, D. (1991); Geohydrology of Aurpalle watershed in semi arid India. Unpub. Ph. D. thesis, Kar. Univ. Dharwad.

Patil, H.G, Deendar, D.I. and Gokhale, N.W. (1984); Mini columnar joints in basalts of Kodali village, Maharashtra. Curr. Sc. Vol. 53, No. 20, p, 1089.

Pujar, G.S. (1989); Geology of the area east of Manoli, Belgaum district, Karnataka state. Unpub, Ph.D. thesis, Kar. Univ. Dharwad.

Puranik, S.C. (1979); Iron formations of Nagavi-Done area, Dharwad district Karnataka state. India. Unpub. Ph. D. thesis Dharwad.

Sederholm, J.J. (1928); On orbicular graites. Comn. Geol. Finland, No. 83, pp. 1-105.

Srikantia S.V., Bhat P.G.K. and Subramani N. (1994); Orbicular rocks from Clospet granite, near Hoskote, Tumkur district, Karnataka. Jour. Geol. Soc. Ind. Vol.43, pp. 267-279.

Tyrrell, G.W. (1980); The principles of petrology. B.I. Publications. New Delhi.

Waghamare, B.P. (1990); Petrological and petrochemical studies of the basic dykes and the associated rocks of Gadag, Kadampur and Nagavi area of Dharwad district, Unpub. Ph.D. thesis, Kar. Univ. Dharwad.